D1476673

URNS

DICE

AND

POLYOMINOES

URNS DICE AND POLYOMINOES

George Barr McCutcheon

ELDERBERRY PRESS, LLC
OAKLAND

Elderberry Press, LLC
1393 Old Homestead Drive, Second floor
Oakland, Oregon 97462—9506.
E-MAIL: editor@elderberrypress. com
elderberrypress. com
TEL/FAX: 541. 459. 6043

All Elderberry books are available from your favorite bookstore, amazon. com, or from or our 24 hour order line: 1. 800. 431. 1579

Library of Congress Control Number: 2003116986
Publisher's Catalog-in-Publication Data
Urns, Dice and Polyominoes / Barr McCutcheon
ISBN 1-930859-85-6
1. Puzzles—— Math.
2. Fun—— Math.
3. Algebra—— Math.
4. Games—— Math.
I. Title

This book was written, printed, and bound in the United States of America.

FOR TOOKIE

Table of Contents

48) How steep is a line?

49) The slope of TU

50) The slope of EA

51) The equation of line KRX

52) Some Geometry

53) πR

54) For centuries...

55) How does the area of the smaller compare...

56) Pythagoras

57) Areas

58) A Day in the Life of a Bird

59) Birds on Holiday

60) If Each of these Marbles....

61) Some rather loud probability

62) Probability with coins

63) Urn J

64) A Person with Three fingers on each hand

65) Suppose we have two fingers on each hand

66) And still fewer

67) Multiplication tables used by our friends

68) Zero is, in some ways, a number...

69) What is zero divided by zero?

70) e

FOREWORD

BY

TOM CAMPBELL

"You must always remember", Barr McCutcheon tells his students, "that everything is twice as big as it would be if it were half as big as it is."

In a day and age when mathematics texts are written with the premise that no one in the classroom, including the teacher, can do much more than follow the given recipe to a solution, Mr. McCutcheon invites his students and readers to come in and marvel at his collection of mathematical ideas and concepts. While many teachers and text book authors feel that they need only explain the facts and the steps to students and the students will then be able to "do math," the sad truth is that it isn't so much the facts of math, but mathematical thinking, that students need to learn how to do. In this book the reader is asked to consider, to see and to understand connections in a way that is both intriguing and exciting. To the question "when are we ever going to have to use this stuff?" which often intimidates mathematics teachers into turning their classrooms into vocational training grounds, Mr. McCutcheon answers, "Well, I am teaching you how to think, and I hope that you will do so every day!" Too few teachers recognize the need to respond to a question with a leading question, helping the students discover the way to their answers. While Mr. McCutcheon is a capable practitioner of this style, he has another role: he is the doorman of the grandest collection of seemingly incredible ideas on earth and he beckons us in to share in them. He doesn't need to ask us questions, but rather to suggest that there is something worthy of our attention involved in the problem at hand, and he then sends us high-tailing off to figure out both the question and the answers, learning and discovering rare gems of knowledge and understanding along the way. In his 49 years of mathematics teaching at the Francis Parker School and Latin School in Chicago, Mr. McCutcheon honed his teaching style as well as the minds of his students. This book is a collection of his educational insights. There are brain-teasers and intellect-ticklers and concepts that stretch well beyond the "useful" facts, into areas that engender deep thought in students of many ages. Mr. McCutcheon does all of this with a sense of humor and classroom presence which comes alive on these pages and which make him one of those memorable teachers with whom former students stay in touch for years.

I had the good fortune to study mathematics with Mr. McCutcheon at Francis Parker School during five of my middle and high school years. I fell in love with both the subject and his style in that time, so much so that I later went back to teach at Parker, under Barr's tutelage, for three years, and am now in my twentieth year of teaching. When Barr asked me several years ago if I would be willing to help him organize his materials for this book, it was like being invited to revisit, and again enjoy, a past pleasure. What it soon became was a renewed opportunity to extend my experience and reinvigorate thoughts and ideas which were not usually part of my daily experience. Recently, we have been trading emails, faxes, overnight packages, and phone calls, and I simply feel lucky to again have had the opportunity to bask in the excitement that he can create about math. On my several short visits to South Carolina, where Barr now lives, we have played with and savored the artistic beauty of this wonderful subject. I am so glad that, although Barr is retired from the classroom, he has continued to teach me about mathematics and I am sure that you will find that through these pages, he will teach you, too. You will also learn, how lucky I and several generations of Mr. McCutcheon's students are, to have been able to experience first-hand his insights and ministrations.

TEC
Portland, ME
December 2003

(1949, 1950, 1951, ······· 1998)

DURING MY FIFTY YEARS OF TEACHING MATHEMATICS IN CHICAGO, AT THE FRANCIS PARKER SCHOOL, THE LATIN SCHOOL, AND ROOSEVELT UNIVERSITY, MY STUDENTS WERE NEVER ASKED TO BUY A TEXTBOOK, OTHER THAN A BOOK OF TABLES. TOWARD THE END OF CLASS, THE HOMEWORK WOULD BE ASSIGNED, EITHER ORALLY, WITH THE BLACKBOARD, OR SOMETIMES BY A GESTURE, SUCH AS POINTING, OR SCRATCHING MY HEAD, OR LOOKING RATHER PUZZLED.

EACH STUDENT WOULD HAVE BEEN ASSIGNED A DIFFERENT Q NUMBER, SO THAT ANY PROBLEM WITH A Q IN IT WOULD BRING, FROM EACH STUDENT, A DIFFERENT ANSWER.

TO MY CLOSE FRIEND AND FORMER STUDENT TOM CAMPBELL,
WHO LIVES WITH HIS WIFE AND THREE CHILDREN
AWAY UP SOMEWHERE ALONG THE SUNSWEPT COAST OF MAINE,
WHO TEACHES MATHEMATICS UP THERE
AND HAS, OVER MANY YEARS, ENCOURAGED ME IN THIS
EFFORT. HE HAS, ESPECIALLY DURING THESE FINAL
MONTHS, ENRICHED HIS COLLABORATIVE OFFERINGS WITH
SYMPATHETIC ADVICE AND HUMOROUS COMPANIONSHIP;
TO MY WIFE TOOKIE, WHO NEVER LOOKS AT BOOKS OR
PAPERS WHICH HAVE ANYTHING TO DO WITH MATHEMATICS
BUT CONTINUES, AFTER A CARD DECK OF YEARS, TO BE
CHEERFUL, LOVING, AND COÖPERATIVE IN ALL OTHER MATTERS;
AND TO OUR SON, IAN ALTIERY McCUTCHEON, WHOSE ARTISTIC
TALENT AND MATHEMATICAL ENTHUSIASM PRODUCED
THIS SPLENDID COVER;

THIS BOOK IS HUMBLY DEDICATED.

CREDO

1) THE PRIMARY PURPOSE OF ANY TEACHER OF MATHEMATICS SHOULD BE TO INSPIRE AS MANY STUDENTS AS POSSIBLE WITH THE FASCINATION OF MATHEMATICAL IDEAS.

2) A SECONDARY PURPOSE SHOULD BE TO DEVELOP IN EVERY STUDENT AS MUCH EFFICIENCY AS POSSIBLE, LESS FOR ITS OWN SAKE THAN TO REMOVE ANY OBSTACLE WHICH MIGHT BECLOUD THE DESIRED FASCINATION.

3) MATHEMATICS IS AN ART, AND TRUE ART CANNOT, BY ITS DEFINITION, BE A TOOL, EXCEPT PARENTHETICALLY, FOR ANYTHING ELSE.

4) THE STUDENT WILL RECEIVE INSPIRATION, IF AT ALL, FROM THE TEACHER AND NOT FROM THE TEXTBOOK.

5) THE PURPOSE OF THE TEXTBOOK, THEREFORE, IS TO HELP THE TEACHER TO CREATE THE PROPER ATMOSPHERE FOR INSPIRATION.

6) A GOOD WAY TO BORE EVERYONE IS TO PRESENT MATHEMATICS IN A HUMORLESS WAY AS A JUMBLE OF DISCOUNTS, COMMISSIONS, AND PILES OF GROCERIES, WITH THE VAGUE PROMISE THAT THE DILIGENT STUDENT WILL ONE DAY BE A SUCCESSFUL BUSINESSMAN OR ENGINEER. (WAS AN INSPIRED LOOK EVER SEEN ON THE FACE OF A STUDENT WHO HAD JUST SUCCEEDED IN COMPUTING A SELLING PRICE?)

7) A TEXTBOOK SHOULD SERVE TO CHANNEL THE THINKING OF THE TEACHER, NOT TO REPLACE IT. THE TEACHER IS URGED TO BECOME FRIENDLY WITH THE IDEAS ON EACH PAGE BEFORE INTRODUCING THEM TO THE CLASS.

IF, IN A CERTAIN YEAR

(NOT A LEAP YEAR)

THE 4TH OF JULY FALLS ON TUESDAY,
ON WHAT DAY OF THE WEEK WILL FALL:

5 JULY?	21 SEPTEMBER
9 JULY?	13 OCTOBER
26 JULY?	25 NOVEMBER
2 AUGUST?	31 DECEMBER
1 SEPTEMBER?	2 FEBRUARY THE FOLLOWING YEAR

IN SOME YEAR, (NOT A LEAP YEAR),
SUPPOSE THAT 1 JANUARY FALLS ON A CERTAIN DAY OF THE WEEK,
SAY Q-DAY

THEN:	2 JANUARY FALLS	ON (Q+1)-DAY
AND	3 JANUARY FALLS	ON (Q+2)-DAY
:		
AND	8 JANUARY WILL FALL	ON (Q+7)-DAY

BUT (Q+7)-DAY IS Q-DAY

SO 15 JANUARY WILL BE	Q-DAY
22 JANUARY WILL BE	Q-DAY
29 JANUARY WILL BE	Q-DAY
31 JANUARY WILL BE	(Q+2)-DAY
1 FEBRUARY WILL BE	(Q+3)-DAY

DO YOU SEE WHY 1 MARCH WILL BE (Q+3)-DAY
AND 1 APRIL WILL BE (Q+6)-DAY
AND 1 MAY WILL BE (Q+8)DAY = (Q+1)-DAY ?

CONTINUING

THIS CHART,

1 JANUARY FALLS ON	Q
1 FEBRUARY	(Q + 3) - DAY
1 MARCH	(Q + 3)
1 APRIL	(Q + 6)
1 MAY	(Q + 1)
1 JUNE	(Q + 4)
1 JULY	(Q + 6)
1 AUGUST	(Q + 2)
1 SEPTEMBER	(Q + 5)
1 OCTOBER	Q
1 NOVEMBER	(Q + 3)
1 DECEMBER	(Q + 5)

S	M	T	W	Th	F	S
1	2	3	4	5	6	7
8	9	10	11	12	13	14
15	16	17	18	19	20	21
22	23	24	25	26	27	28
29	30	31	32	33	. . .	
36	37	38	. . . .			
43	. . . .					
. . . .						

TO FIND THE DAY OF THE WEEK
FROM 1 MARCH, 2000
THROUGH 2099
WE SIMPLY ADD

1) THE DATE NUMBER: 1, 2, 3, ··· 31
2) THE MONTH NUMBER
3) THE LAST 2-DIGIT NUMBER OF THE YEAR
(SINCE $365 = 52(7) + 1$,
EACH YEAR THE DATE INCREASES BY 1
THE DAY, ALSO, INCREASES BY 1)
4) THE NUMBER OF PRECEDING 29 FEBRUARIES, STARTING WITH THE ONE IN 2004

EXAMPLES:

	DATE	MONTH	YEAR	29 FEBS	
5 JUNE, 2002	5	4	2	0	WED.
23 AUGUST, 2009	23	2	9	2	SUN.
19 OCTOBER, 2027	19	0	27	6	TUES.
15 APRIL, 2079	15	6	79	19	SAT.

CALENDAR
FOR THE
21ST
CENTURY

J	F	M	A	M	J	J	A	S	O	N	D
0	3	3	6	1	4	6	2	5	0	3	5

0 1 2 3 4 5 6 7

8 12

16 20

24 28

32

36 40

44 48

52 56

60

64 68

72 76

80 84

88

92 96

11	11
APRIL	6
2017	17
L.Y.	4
	38 TUES.

9	9
JULY	6
2008	8
L.Y.	2
	25 WED.

20	20
NOV.	3
2043	43
L.Y.	10
	76 FRI.

15	15
OCT.	0
2024	24
L.Y.	6
	45 TUES.

25	25
JAN	0
2032	32
L.Y.	7
	64 SUN.

23	23
FEB.	3
2044	44
L.Y.	10
	80 TUES.

31	31
AUG	2
2077	77
	19
	129 TUES.

25	25
DEC.	5
2083	83
L.Y.	20
	133 SAT.

S	M	T	W	Th	F	S
50	51	52	53	54	55	56
43	44	45	46	47	48	49
36	37	38	39	40	41	42
29	30	31	32	33	34	35
22	23	24	25	26	27	28
15	16	17	18	19	20	21
8	9	10	11	12	13	14
1	2	3	4	5	6	7

FIND DAY OF WEEK

1) 13 MARCH, 2005

2) 17 SEPTEMBER, 2016

3) 31 JANUARY, 2054

4) 9 DECEMBER, 2085

5) 11 FEBRUARY, 2032

6). IN WHICH YEARS DURING THIS CENTURY WILL APRIL FOOLS DAY FALL ON SATURDAY ?

7) DURING WHICH YEARS IN THIS CENTURY WILL 12 FEBRUARY AND 11 MARCH BOTH FALL ON WEDNESDAY ?

ESTIMATION

WHEN WE ADD 2 + 3, IT IS NOT NECESSARY TO ESTIMATE, BECAUSE WE ALREADY HAVE A FAIRLY GOOD IDEA OF THE ANSWER.

HOWEVER, IF WE WANT TO MULTIPLY 7.8943 BY $10\frac{11}{17}$, WHICH OF THE FOLLOWING IS CLOSEST TO THE ANSWER?

18 70 250 88 1000

A GOOD GUESS FOR $(9.8376)^4$ IS ☐

ADDING $19\frac{3}{5}$, 20.3, $19\frac{7}{9}$, 20.1487, AND 21

GIVES US APPROXIMATELY ☐

IS ONE OF THE NUMBERS BELOW MUCH BIGGER OR MUCH SMALLER THAN THE OTHERS?

$(23.78)^2$, 247 TIMES 2.1, 3^8, $\frac{1111}{2}$, 7 TIMES 8 TIMES 9

WHAT ABOUT THESE?

$(1.8)^5$, $\frac{55}{3}$, 6.9 TIMES 2.9, $(.89)^7$, $(4.3)^2$

ADD THESE IN 10 SECONDS OR LESS:

49 + 51 + 47 + 53 + 52 + 48 + 70 + 30 + 50 + 43

WHAT, APPROXIMATELY, IS THE SQUARE ROOT OF 888.8?

ESTIMATE ANSWERS FOR THE FOLLOWING:

1) 12 TIMES 68 =

2) $15.5 + 33\frac{1}{3} =$

3) $7^4 =$

4) 18.9 TIMES 1.89 =

5) $24 + 26 + 23 + 27 + 19 + 31 + 23 + 10 =$

6) $\sqrt{87}$

7) $(2.1)^5$

8) $\sqrt{1555}$

9) $\sqrt{6399}$

10) $(1.01)^{10}$

11) $\sqrt{63.978}$

12) $(99.87)(100.1)$

ESTIMATION

WHICH OF THE FOLLOWING NUMBERS IS SMALLEST?

$\left(\frac{1}{2}\right)^{10}$ 2 TIMES 10^{-3} $(.1)^{[(2^2)^2]}$

$\sqrt{.01}$ $\left(\frac{1}{3}\right)^{10}$ $\frac{1}{1000000}$

WHICH OF THE FOLLOWING NUMBERS IS LARGEST?

10^{10} 12^{8} 6^{10}

$3^{(3^2)}$ 9^{5} $2^{[(2^2)^2]}$

WHAT IS $(10.97)^3$?

WHAT IS $\sqrt{63.9732}$?

$\frac{1}{2}$ IS A FRACTION. WE HAVE SOME SENSE OF WHAT ONE HALF MEANS ....

IF $\frac{1}{2}$ OF THE PEOPLE IN THIS ROOM ARE DOG OWNERS, WE CAN PICTURE A ROOM WHERE THAT IS SO.

A SKETCH OF ONE HALF, $(\frac{1}{2})$, MIGHT LOOK LIKE THIS

OR THIS

THERE ARE MANY OTHER FRACTIONS, SUCH AS $\frac{3}{4}$, $\frac{2}{5}$, $\frac{1}{7}$, AND $\frac{112}{243}$.

$\frac{2}{3}$ CAN BE DRAWN

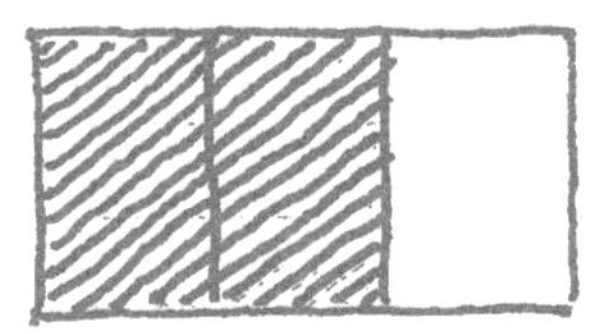

$\frac{4}{6}$ A ——— B

ERASING LINE AB GIVES US THE SKETCH OF $\frac{2}{3}$.

HERE IS $\frac{6}{9}$

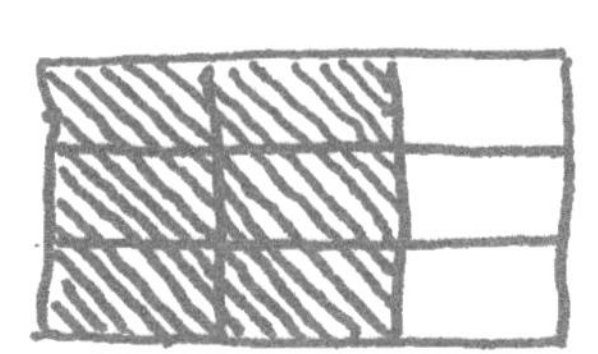

ERASING HORIZONTAL LINES GIVES US $\frac{2}{3}$.

HENCE

$\frac{2}{3} = \frac{4}{6} = \frac{6}{9}$.

$>$ MEANS BIGGER THAN

$<$ MEANS LESS THAN

$=$ MEANS EQUAL TO

$\frac{1}{2} = \frac{4}{8}$ $\qquad$ $\frac{1}{2} = \frac{3}{\;}$ $\qquad$ $\frac{1}{2} = \frac{\;}{10}$ $\qquad$ $\frac{1}{2} = \frac{W}{\;}$

$\frac{2}{5} = \frac{6}{15}$ $\qquad$ $\frac{10J}{15J} = \frac{2}{3}$ $\qquad$ $\frac{2M}{3M} = \frac{6}{\;}$

$\frac{44Y}{55Y} = \frac{4}{\;}$ $\qquad$ $\frac{14M}{\;} = \frac{2}{3}$ $\qquad$ $\frac{2}{5} = \frac{20}{\;}$ $\qquad$ $\frac{2}{3} = \frac{8K}{\;}$

$\frac{9}{15} = \frac{\;}{50}$ $\qquad$ $\frac{\;}{20K} = \frac{3}{4}$ $\qquad$ $\frac{4}{7} = \frac{36}{\;}$

$\frac{2}{3} \quad \frac{3}{4}$ $\qquad$ $\frac{5}{6} \quad \frac{9}{12}$ $\qquad$ $\frac{3W}{5W} \quad \frac{6}{10}$ $\qquad$ $\frac{4}{9} \quad \frac{7}{18}$

1) $\frac{1}{2} + \frac{1}{4}$

$\frac{2}{4} + \frac{1}{4} = \boxed{\frac{3}{4}}$

2) $\frac{2}{3} + \frac{1}{5}$

3) $\frac{1}{3} + \frac{3}{4}$

4) $2\frac{2}{3} + 5\frac{1}{5}$

5) $2\frac{3}{5} + 7\frac{7}{10}$

6) $6\frac{1}{4} + 2\frac{5}{6}$

7) $\frac{5}{W} + \frac{7}{W}$

8) $\frac{3}{J} + \frac{5}{2J}$

9) $\frac{3}{\Diamond} + \frac{4}{3\Diamond}$

10) $8\frac{3}{5} - 2\frac{1}{5}$

11) $4\frac{2}{3} - 1\frac{1}{6}$

12) $6\frac{5}{7} + 1\frac{3}{14}$

13) $5\frac{1}{8} - 2\frac{3}{4}$

14) $4\frac{1}{3} - 3\frac{5}{6}$

15) $5\frac{2}{3} + 2\frac{5}{6}$

16) $4\frac{3}{11} + 5\frac{5}{7}$

1) $\dfrac{3\frac{1}{2}}{5\frac{1}{4}} =$

2) $4 + 2\frac{2}{3} =$

3) $\frac{4}{5} \div \frac{3}{7} =$

4) $2\frac{3}{5} \div 5\frac{1}{4}$

5) $6\frac{1}{3} \div 3\frac{4}{5}$

6) $4\frac{1}{7} + \frac{3}{M}$

7) $\frac{K}{7} + \frac{2}{3R} + 2\frac{1}{2}$

8) $\frac{\pi}{3} + 4\frac{2}{5}$

9) $\frac{J}{5} + \frac{T}{3} + \frac{M}{7}$

10) $\frac{1}{A} + \frac{1}{B} + \frac{1}{C} =$

11) $W + 3\frac{2}{5} = 8$

$W =$

12) $B + 4\frac{1}{2} = 11\frac{3}{8}$

$B =$

13) $T + 2T = 4\frac{3}{5}$

14) $3\frac{2}{3}K = 8\frac{1}{5}$

EQUATIONS

1) $J + 3 = 7$

WHAT IS THE VALUE OF J?

J MUST EQUAL 4.

$\boxed{J = 4}$

2) $D + D = 9$

NO WHOLE NUMBER WILL WORK.

D MUST EQUAL $4\frac{1}{2}$.

$\boxed{D = 4\frac{1}{2}}$

3) $W + 7 = 20$

W MUST EQUAL 13

$\boxed{W = 13}$

4) $Q + 8 = 3Q + 1$

WE DRAW A PICTURE

FROM EACH SIDE

WE REMOVE A Q.

FROM EACH SIDE

WE REMOVE 1

$7 = Q + Q$

$2Q = 7$

$\boxed{Q = 3\frac{1}{2}}$

5) $9M + 2 = 4M + 8$

(HINT: FIRST REMOVE

4M FROM EACH SIDE.)

WELL,
THAT WASN'T
VERY HARD!

WHAT ABOUT
$3K = 21$
(OR $K+K+K=21$)

WHAT VALUE OF K
WILL MAKE THESE TRAYS
BALANCE?

7? 777 21 YES.

SO, $K = 7$

$2w + 11 = 5w + 5$

WE CAN REMOVE
FROM EACH TRAY
TWO OF THESE W'S

11 | 5 w w w

$11 = w + w + w + 5$

$6 = 3w$

$2 = w$

$M + 5 = 8$

ANSWER
$M = 3$

$V + V + 7 = 12$

$V =$

$C + 5 = 0$

$C =$

$7\# + 7 = \# + 25$

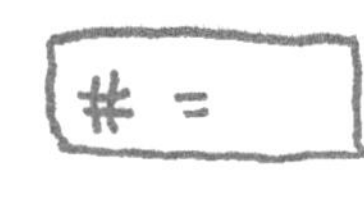

$5 + 6 = J + J$

$L + L = L$

$L =$

$4B + 10 = 23$

$B =$

$3D + 5 = 5D + 2$

$= D$

$t + t + t = 17$

$t =$

$2Z + 22 = 5Z$

$= Z$

$117 = 5s + 5s + 5s$

$= s$

$5u + u = 47$

$5T + 2 = 56$

$T =$

$P + P = 0$

$P =$

$3R = 1$

$R =$

$5\Delta - 2 = 30 - 3\Delta$

$K + 2 = 11$

$K =$ ☐

$@ + 7 = 13$

$@ =$ ☐

1) $M + 17 = 22$

2) $3Q = 21$

3) $T - 8 = 31$

4) $U + 4U = 1$

5) $J + J = 19$

6) $5V = 2$

7) $P + 7 = P$

8) $\frac{W}{3} = 4\frac{2}{5}$

9) $2W + 7 = 5W + 5$

10) $\frac{3}{J} = 4\frac{2}{5}$

11) $4M + 7 = 9M + 3$

12) $\frac{3T}{5} = \frac{10}{T}$

13) $\frac{U}{4} = 3\frac{2}{5}$

14) $L - 3 = 16 - 2L$

15) $\Delta + 4\Delta + 7 = 0$

16) $3* + 2 = * + 27$

1) ARCHIE WEIGHS TWICE AS MUCH AS BASIL. BASIL WEIGHS TWICE AS MUCH AS CHARLIE. ALL TOGETHER THE THREE WEIGH 182 POUNDS. FIND ARCHIE'S WEIGHT.

2) SALLY HAS THREE TIMES AS MANY PENNIES AS AGNES. EVY HAS 3 TIMES AS MANY AS SALLY. ALL TOGETHER THEY HAVE $3.77 HOW MANY PENNIES HAS EVY?

3) WILVO HAS TWICE AS MANY ATLATLS AS TURNER. WILVO TAKES 5 ATLATLS FROM TURNER, GIVES ONE AWAY, AND ENDS UP WITH 3 TIMES AS MANY AS TURNER. HOW MANY DID WILVO HAVE AT FIRST?

4) ISAIAH IS 3 TIMES AS OLD AS HEZEKIAH. IN 4 YEARS ISAIAH WILL BE TWICE AS OLD AS HEZEKIAH. HOW OLD WILL HEZEKIAH BE, WHEN ISAIAH IS 24 YEARS OLD?

5) PIMBOWL HAS 4 TIMES AS MANY WIDGETS AS ROCKJAW. ROCKJAW HAS 1 MORE THAN 4 TIMES AS MANY WIDGETS AS SAWDUKE. IF THE 3 HAVE 89 WIDGETS TOGETHER, HOW MANY WIDGETS HAS PIMBOWL?

6) ALL TOGETHER RAJEEV, SHAHAN, AND TENGAT HAVE $20.95 WORTH OF NICKELS. IF RAJEEV HAS HALF AS MANY AS SHAHAN, AND TENGAT HAS 1 LESS THAN HALF AS MANY AS RAJEEV, HOW MANY NICKELS HAS RAJEEV?

7) JUDY HAS 3 TIMES AS MANY ATLATLS AS GAIL. GAIL HAS 1 MORE THAN TWICE AS MANY ATLATLS AS CATHY. IF THE THREE HAVE 643 ATLATLS BETWEEN THEM, HOW MANY HAS GAIL?

8) PAUL HAS $6.99 IN NICKELS, DIMES, AND PENNIES. HE HAS ONE MORE NICKEL THAN HE HAS PENNIES AND ONE MORE PENNY THAN HE HAS DIMES. HOW MANY DIMES HAS PAUL?

THE THREE IVITIES

4 + 5 = 9
5 + 4 = 9
4 + 5 = 5 + 4

7 + 12 = 19
12 + 7 = 19

23 + 75 = 75 + 23

ADDITION IS COMMUTATIVE.

8 TIMES 3 = 24
3 TIMES 8 = 24
8 TIMES 3 = 3 TIMES 8

9 TIMES 6 = 54
6 TIMES 9 = 54

21 TIMES 7 = 7 TIMES 21

MULTIPLICATION IS COMMUTATIVE.

7 - 4 = 3
4 - 7 = -3
7 - 4 DOES NOT EQUAL 4 - 7.

11 - 6 = 5
6 - 11 = -5

SUBTRACTION IS NOT COMMUTATIVE.

THE AVERAGE OF 6 AND 9 IS $7\frac{1}{2}$.

THE AVERAGE OF 9 AND 6 IS $7\frac{1}{2}$.

FINDING THE AVERAGE IS COMMUTATIVE.

3 TO THE POWER 4 IS $3^4 = 81$

4 TO THE POWER 3 IS $4^3 = 64$

$2^{10} = 1024$ $10^2 = 100$

RAISING TO A POWER IS NOT COMMUTATIVE.

10 DIVIDED BY 2 IS 5
2 DIVIDED BY 10 IS $\frac{1}{5}$

DIVISION IS NOT COMMUTATIVE.

$3+(4+8) = 3+12 = 15$
$(3+4)+8 = 7+8 = 15$
SO, $3+(4+8) = (3+4)+8$
AND ADDITION IS ASSOCIATIVE

$8-(7-3) = 8-4 = 4$
$(8-7)-3 = 1-3 = -2$
SO, $(8-7)-3$ DOES NOT EQUAL $8-(7-3)$

AND SUBTRACTION IS NOT ASSOCIATIVE

12 DIVIDED BY (6 DIVIDED BY 2) = 12 DIVIDED BY 3 = 4
(12 DIVIDED BY 6) DIVIDED BY 2 = 2 DIVIDED BY 2 = 1
SO, 12 DIVIDED BY (6 DIVIDED BY 2) DOES NOT EQUAL (12 DIVIDED BY 6) DIVIDED BY 2

AND DIVISION IS NOT ASSOCIATIVE.

(5 TIMES 7) TIMES 3 = 35 TIMES 3 = 105
5 TIMES (7 TIMES 3) = 5 TIMES 21 = 105
SO, (5 TIMES 7) TIMES 3 DOES EQUAL 5 TIMES (7 TIMES 3)

AND MULTIPLICATION IS ASSOCIATIVE.

4 AVERAGE (8 AVERAGE 20) = 4 AVERAGE 14 = 9
(4 AVERAGE 8) AVERAGE 20 = 6 AVERAGE 20 = 13

SO, 4 AVERAGE (8 AVERAGE 20) DOES NOT EQUAL (4 AVERAGE 8) AVERAGE 20

AND FINDING THE AVERAGE IS NOT ASSOCIATIVE.

$(2^3)^4 = 8^4 = 4096$ (WHICH IS ONLY 2^{12})

$2^{(3^4)} = 2^{81}$

SO, RAISING TO A POWER IS NOT ASSOCIATIVE.

4 TIMES (3 PLUS 2) =
(4 TIMES 3) PLUS (4 TIMES 2)

4 TIMES 5 = 20
12 PLUS 8 = 20

SO, TIMES DISTRIBUTES OVER PLUS.

7 TIMES (5 MINUS 3)
(7 TIMES 5) MINUS (7 TIMES 3)

7 TIMES 2 = 14
35 MINUS 21 = 14

SO, TIMES DISTRIBUTES OVER MINUS

DOES PLUS DISTRIBUTE OVER TIMES?
DOES 3 PLUS (6 TIMES 7) EQUAL (3 PLUS 6) TIMES (3 PLUS 7)?

DOES (3 PLUS 42) = 9 TIMES 10 NO.
SO PLUS DOES <u>NOT</u> DISTRIBUTE OVER TIMES.

ATLATLS

SAM HAS 5 TIMES AS MANY ATLATLS AS JOE.
THEN SAM GIVES JOE 7 ATLATLS.
THIS LEAVES SAM WITH ONLY 4 TIMES AS MANY ATLATLS AS JOE.
HOW MANY ATLATLS HAVE THEY ALL TOGETHER?

FIRST WE MAKE A GUESS.
LET'S SAY THAT AT FIRST JOE HAD 20 ATLATLS.
THEN AT FIRST SAM MUST HAVE HAD 100 ATLATLS.
AFTER THE GIFT JOE 27
SAM 93. IF OUR GUESS IS CORRECT,
THEN 93 WILL BE 4 TIMES 27.
OOPS, WE GUESSED WRONG.

AFTER THIS, WE MAKE ANOTHER GUESS:
LET'S SAY THAT JOE STARTED WITH 30 ATLATLS.
THEN SAM MUST HAVE STARTED WITH 150 ATLATLS.
AFTER THE GIFT JOE 37 THEN 143 SHOULD BE 4 TIMES 37.
SAM 143 CLOSER, BUT STILL NOT RIGHT.

AFTER TWO WRONG GUESSES, WE WANT TO MAKE A CORRECT GUESS.
BUT SINCE WE DON'T KNOW WHAT THE CORRECT GUESS IS,
WE CALL IT W AND GUESS W.
JOE AT FIRST W
SAM AT FIRST 5W
AFTER THE GIFT JOE W+7
SAM 5W−7
THEN SINCE OUR GUESS IS CORRECT
5W−7 WILL CERTAINLY BE 4 TIMES W+7.
SO WE WRITE $5W-7=4(W+7)$
OR $5W-7=4W+28$
SUBTRACTING 4W FROM EACH SIDE, $W-7=28$
$W=35$

DOES W=35 WORK? JOE AT FIRST 35
SAM AT FIRST 175
AFTER THE GIFT JOE 42, SAM 168
DOES 168 EQUAL 4 TIMES 42? IT DOES.

TRAFFIC SIGNALS

$4+3$ MEANS THAT 4 AND 3 ARE TO BE ADDED.

43 MEANS SOMETHING QUITE DIFFERENT.
IT MEANS 4 TENS PLUS 3, OR $40+3$.

WITH LETTERS, THE CONVENTIONS ARE DIFFERENT.

$J+K$ MEANS THAT J AND K ARE TO BE ADDED.
AND IF WE DON'T KNOW WHICH NUMBERS
J AND K STAND FOR, WE LEAVE IT AS $J+K$.

JK, HOWEVER, MEANS THAT J AND K ARE TO BE MULTIPLIED,
AND IF WE DON'T KNOW WHICH NUMBERS J AND K STAND FOR,
WE LEAVE IT AS JK.

IF ONE OF THE SYMBOLS IS A NUMBER AND THE OTHER A LETTER,
THE MEANING IS THE SAME AS IF BOTH ARE LETTERS:
$3+M$, OR $M+3$, MEANS ADD 3 AND M.
$3M$ MEANS MULTIPLY 3 TIMES M.
(NOTE: M3 IS NEVER WRITTEN, SINCE IT
WOULD BE EASILY CONFUSED WITH M^3, WHICH
MEANS M TIMES M TIMES M.)

POSSIBLY BECAUSE THE 3 IS CLOSER TO THE M IN MULTIPLICATION ($3M$)
THAN IT IS IN ADDITION ($3+M$)
IT HAS BECOME CONVENTIONAL
FOR MULTIPLICATION TO TAKE PRECEDENCE OVER ADDITION,
SO THAT $(2)(3)+(4)(5) = 6+20 = 26$
$2+(3)(4)+5 = 2+12+5 = 19$
AND $2+3+(4)(5) = 2+3+20 = 25$

ONLY ONE OF THESE STATEMENTS IS TRUE.

$3 + 2 + 5 = 10$
$3 + 2 + 5 = 11$
$3 + 2 + 5 = 13$
$3 + 2 + 5 = 21$
$3 + 2 + 5 = 25$
$3 + 2 + 5 = 30$

HERE ARE FOURTEEN PARENTHESES:

CAN YOU PUT THESE FOURTEEN PARENTHESES INTO FIVE OF THE SIX STATEMENTS IN SUCH A WAY THAT ALL SIX STATEMENTS WILL BE TRUE?

WHAT DOES THIS MEAN?

$(3 + (2)5)(4 + 1)$

IT MEANS

$(3 + 10)(5)$

OR

$(13)(5) = 65$

BUT NESTED PARENTHESES CAN BE CONFUSING.

SO IT IS PREFERABLE TO WRITE

$[3 + (2)5](4 + 1)$,

REPLACING THE OUTER PARENTHESES WITH BRACKETS.

$7 + 3 =$
$7(+3) =$
$4 + 6 + 3 =$
$4 + 6(+3) =$
$4(+6) + 3 =$
$4(+6 + 3) =$
$(4 + 6)(+3) =$
$5 + 7 + 2 =$
$5(+7) + 2 =$
$5 + 7(+2) =$
$5(+7)(+2) =$
$5(+7 + 2) =$
$(5 + 7)(+2) =$

$7(-3) =$
$7 - 3 + 2 =$
$7 - (3 + 2) =$
$-4 - 3 =$
$-(4 - 3) =$
$(-4)(-3) =$
$4 - 5 + 7 =$
$4(-5 + 7) =$
$4(-5) + 7 =$
$4 - (5 + 7) =$
$4 - 5(+7) =$
$4(-5)(+7) =$

IF $w = 3$ AND $v = 2$

FIND $3w^2 + 4v^2$

$3(w^2 + 4v^2)$

$3(w^2 + 4v)^2$

$3[w^2 + (4v)^2]^2$

$3[(w^2 + 4v)^2 + 5w]^2$

MANY OF THESE EXPRESSIONS
HAVE MEANING.

BUT SOME DO NOT. CAN YOU EVALUATE THE EXPRESSIONS WHICH HAVE MEANING?

1) $4(-)3-2$

2) $4(-3)-2$

3) $4(-3-)2$

4) $4(-3-2)$

5) $4-(3)-2$

6) $4-(3-)2$

7) $4-(3-2)$

8) $4-3(-)2$

9) $4-3(-2)$

10) $4-3-(2)$

11) $(4-3)(-2)$

12) $4(-3)(-2)$

13) $4+23$

14) $4+(2)3$

15) $4(+23)$

16) $4(+2)3$

17) $(4+2)3$

18) $4(+2)3$

19) $4+(23)$

20) $4(+)23$

21) $(4+)(2)(3)$

22) $(4+(2)3$

23) $(4)+(2)(3)$

24) $[(4)+(2)(3)]$

25) $4)-(3-2$

26) $4(+2(3$

27) $4+]2[3$

WHERE CAN WE DEPLOY PARENTHESES

TO MAKE THESE STATEMENTS TRUE?

1) 3 + 4 + 5 = 17

2) 3 + 4 + 5 = 23

3) 3 + 4 + 5 = 27

4) 3 + 4 + 5 = 35

5) 3 + 4 + 5 = 60

6) 2 3 + 4 5 = 26

7) 2 3 + 4 5 = 46

8) 2 3 + 4 5 = 50

9) 2 3 + 4 5 = 51

10) 2 3 + 4 5 = 70

11) 2 3 + 4 5 = 96

12) 2 3 + 4 5 = 120

13) 2 3 + 4 5 = 270

14) 2 3 + 4 5 = 460

15) 2 3 + 4 5 = 1035

BARBARA IS 6 TIMES AS OLD AS AMY.
5 YEARS AGO BARBARA WAS 8 TIMES AS OLD AS AMY.
HOW OLD IS BARBARA NOW?

[HINT: AMY IS YOUNGER. MAKE A GUESS AT AMY'S AGE NOW.
IF YOU GUESS A NUMBER, SEE IF IT WORKS.
IF IT DOESN'T WORK, DON'T WORRY.
NOW MAKE A CORRECT GUESS (A LETTER)
AS IN THE PROBLEM ABOVE.]

ALASDAIR IS 4 TIMES AS OLD AS ANNIE.
7 YEARS AGO ALASDAIR WAS 5 TIMES AS OLD AS ANNIE.
HOW OLD WAS ALASDAIR WHEN ANNIE WAS BORN?
[HINT: ALASDAIR IS PRETTY WELL PAST MIDDLE AGE]

1) IF ELSA WAS 5 YEARS OLD WHEN FRANK WAS BORN, HOW OLD WILL FRANK BE WHEN ELSA IS 22?

2) IF HILDEGARDE HAS 7 JUMPROPES, AND SHE AND HER SISTER INGA RECEIVE A NEW JUMPROPE EACH, THEN HILDEGARDE HAS TWICE AS MANY JUMPROPES AS INGA. HOW MANY JUMPROPES HAD INGA AT FIRST?

3) SAM HAS TWICE AS MANY ATLATLS AS TRAN. RICARDO HAS TWICE AS MANY AS SAM. ALL TOGETHER THEY HAVE 91 ATLATLS. HOW MANY HAS RICARDO?

4) DAGS IS 11 YEARS OLD AND WURFLE IS 2. WILL THERE BE A TIME WHEN DAGS WILL BE TWICE AS OLD AS WURFLE? IF SO, WHEN?

5) ADDLE HAS 3 TIMES AS MANY PENNIES AS BURL. AFTER ADDLE GIVES BURL 7 PENNIES, ADDLE WILL HAVE ONLY TWICE AS MANY AS BURL. HOW MANY HAVE THEY ALL TOGETHER?

6) GUIDO HAS 8 TIMES AS MANY ATLATLS AS HIS FRIEND PAULIE. GUIDO GIVES PAULIE 30 ATLATLS FROM HIS COLLECTION. GUIDO NOW HAS ONLY TWICE AS MANY ATLATLS AS PAULIE. HOW MANY ATLATLS HAD GUIDO TO BEGIN WITH?

POWERS

TWO TIMES TWO EQUALS FOUR
CAN BE WRITTEN AS

$2 \times 2 = 4$, OR AS $2^2 = 4$

TWO TIMES TWO TIMES TWO EQUALS EIGHT
CAN BE WRITTEN AS

$2 \times 2 \times 2 = 8$, OR AS $2^3 = 8$

THREE TIMES THREE EQUALS NINE
CAN BE WRITTEN AS

$3 \times 3 = 9$, OR AS $3^2 = 9$

FIVE TIMES FIVE TIMES FIVE EQUALS ____
CAN BE WRITTEN AS

________ OR AS _______

$7^2 =$ ____ $(2+4)^{1+2} =$ ____

$2^7 =$ ____ $(7+3)^{2+4} =$ ____

1) $2^3 \cdot 2^4 = 2^{\square}$

2) $7^4 \cdot 7^5 = 7^{\square}$

3) $Q^5 \cdot Q^7 = Q^{\square}$

4) $X^3 \cdot X^4 \cdot X^5 = X^{\square}$

5) $X^A \cdot X^B = X^{\square}$

6) $(J^3)^2 = J^{\square}$

7) $(M^4)^5 = M^{\square}$

8) $(T^C)^D = T^{\square}$

9) $36^{\frac{1}{2}} =$

10) $27^{\frac{1}{3}} =$

POWERS

7 TIMES 7 IS SOMETIMES WRITTEN 7^2,
AND IT EQUALS 49.

5 TIMES 5 TIMES 5 CAN BE WRITTEN 5^3
AND IT EQUALS 125.

3 TIMES 3 TIMES 3 TIMES 3 = 3^4 = 81

2 TIMES 2 TIMES 2 TIMES 2 TIMES 2
TIMES
2 TIMES 2 TIMES 2 TIMES 2
CAN BE WRITTEN
2^5 TIMES 2^4
WHICH IS
32 TIMES 16
(DOES THE ANSWER APPEAR ON THIS PAGE ?)

WHAT IS 2^6 TIMES 2^{16} ?

WHAT IS 128 TIMES 256 ?

WHAT IS 3^4 TIMES 3^7 ?

WHAT IS 243 TIMES 729 ?

WHAT IS 5^3 TIMES 5^4 ?

WHAT IS 3125 TIMES 15625 ?

$2^1 = 2$
$2^2 = 4$
$2^3 = 8$
$2^4 = 16$
$2^5 = 32$
$2^6 = 64$
$2^7 = 128$
$2^8 = 256$
$2^9 = 512$
$2^{10} = 1024$
$2^{11} = 2048$
$2^{12} = 4096$
$2^{13} = 8192$
$2^{14} = 16384$
$2^{15} = 32768$
$2^{16} = 65536$
$2^{17} = 131072$
$2^{18} = 262144$
$2^{19} = 524288$
$2^{20} = 1048576$
. . . . .
$2^{23} = 8388608$
. . . . .
$2^{30} = 1073741824$
. . . . .

$3^1 = 3$
$3^2 = 9$
$3^3 = 27$
$3^4 = 81$
$3^5 = 243$
$3^6 = 729$
$3^7 = 2187$
$3^8 = 6561$
$3^9 = 19683$
$3^{10} = 59049$
$3^{11} = 177147$
. . . . . .

5
25
125
625
3125
15625
78125
390625
1953125
9765625
48828125
. . . .

THESE TINY NUMBERS,

ABOVE, AND TO THE RIGHT
OF THE LARGER NUMBER,
ARE CALLED
POWERS
OR
EXPONENTS.

WHAT HAPPENS IF WE EXTEND UPWARD
EACH OF THESE POWER TABLES ?

. . . .	. . . .	. . . .
$2^{-2} = \frac{1}{4}$	$3^{-2} = \frac{1}{9}$	$5^{-2} = \frac{1}{25}$
$2^{-1} = \frac{1}{2}$	$3^{-1} = \frac{1}{3}$	$5^{-1} = \frac{1}{5}$
$2^{0} = 1$	$3^{0} = 1$	$5^{0} = 1$
$2^{1} = 2$	$3^{1} = 3$	$5^{1} = 5$
$2^{2} = 4$	$3^{2} = 9$	$5^{2} = 25$

OF COURSE, IT IS NOT CLEAR
WHAT 3^{-2} REALLY MEANS, IF ANYTHING.
BUT THE BEHAVIOR OF THE EXPONENTS ON THESE PAGES
IS SO SIMPLE AND CHEERFUL
THAT WE ASSIGN MEANINGS AND VALUES
TO 5^{0}, 3^{-1}, AND THE REST,
SO THAT THIS BEHAVIOR WILL CONTINUE :

SO THAT 5^{0} TIMES 5^{2} WILL EQUAL 5^{2}
AND SO THAT 3^{-1} TIMES 3^{2} WILL EQUAL 3^{1}.

WHAT SHALL $9^{\frac{1}{2}}$ MEAN ?
SURELY NOT ς OR ∃ !

SINCE WE LIKE SO MUCH THE BEHAVIOR OF THESE EXPONENTS,
WE SHALL INSIST THAT $9^{\frac{1}{2}}$ TIMES $9^{\frac{1}{2}}$ SHALL EQUAL 9^{1}, WHICH IS 9,
AND SINCE $9^{\frac{1}{2}}$ TIMES ITSELF EQUALS 9,
$9^{\frac{1}{2}}$ MUST EQUAL 3.

POWERS

ON THE PREVIOUS PAGE WE FOUND THAT

$$2^5 = 2 \times 2 \times 2 \times 2 \times 2 = 32, \quad 7^0 = 1, \text{ AND } 25^{\frac{1}{2}} = 5$$

EVALUATE THE FOLLOWING:

1) $2^3 \cdot 2^4$

2) $7^4 \cdot 7^5$

3) $a^5 \cdot a^7$

4) $x^3 \cdot x^4 \cdot x^5$

5) $(J^3)^7$

6) $M^A M^B$

7) $(T^C)^D$

8) $(2^3)^4$

9) $(4^3)^2$

10) $2^{(3^4)}$

11) $4^{(3^2)}$

12) $9^{\frac{1}{2}}$

13) $\frac{5^7}{5^3}$

POTATOES

1) IF 3 IDENTICAL BILLIARD BALLS WEIGH 19 OUNCES,
HOW MUCH WOULD 13 OF THEM WEIGH?

2) IF 3 SERGEANTS CAN PEEL A BIN OF POTATOES IN 5 HOURS
HOW LONG WILL IT TAKE 1 SERGEANT
TO PEEL 2 BINS OF POTATOES?
HOW LONG WILL IT TAKE 2 SERGEANTS
TO PEEL A BIN OF POTATOES?

3) IF 3 COWS CAN EAT 5 BALES OF HAY IN 8 DAYS,
HOW LONG WILL IT TAKE 5 COWS TO EAT 3 BALES?

4) IF 2 OF ZEKE'S PLUMBERS CAN FIX 5 LEAKS IN 3 WEEKS,
HOW LONG WILL IT TAKE 3 OF THEM TO FIX 7 LEAKS?

5) IF SAM CAN WALK 600 FEET IN 5 MINUTES
AND RUN 600 FEET IN 2 MINUTES,
HOW LONG WOULD IT TAKE HIM TO GO THE 600 FEET
IF HE RUNS FOR 225 FEET AND THEN
WALKS THE REST OF THE WAY?

6) ONE WATER HOSE CAN FILL A POOL IN 8 HOURS.
ANOTHER HOSE CAN FILL THE POOL IN 20 HOURS.
IF BOTH HOSES ARE TURNED ON,
HOW LONG WILL THEY TAKE TO FILL THE POOL?

WHAT DO WE DO WITH AN EQUATION WHICH HAS AN X^2 IN IT ?!@

TAKE

$$AX^2 + BX + C = 0$$

WE DIVIDE BOTH SIDES BY A.

$$X^2 + \frac{B}{A}X + \frac{C}{A} = 0$$

$$X^2 + \frac{B}{A}X + \frac{B^2}{4A^2} - \left(\frac{B^2}{4A^2} - \frac{C}{A}\right) = 0$$

$$X^2 + \frac{B}{A}X + \frac{B^2}{4A^2} - \left(\frac{B^2}{4A^2} - \frac{4AC}{4A^2}\right) = 0$$

$$\left(X + \frac{B}{2A}\right)^2 - \frac{B^2 - 4AC}{4A^2} = 0$$

$$\left(X + \frac{B}{2A}\right)^2 - \left(\frac{\sqrt{B^2 - 4AC}}{2A}\right)^2 = 0$$

$$\left[\left(X + \frac{B}{2A}\right) + \frac{\sqrt{B^2 - 4AC}}{2A}\right]\left[\left(X + \frac{B}{2A}\right) - \frac{\sqrt{B^2 - 4AC}}{2A}\right] = 0$$

SO, EITHER $X + \frac{B}{2A} + \frac{\sqrt{B^2 - 4AC}}{2A} = 0$

OR

$$X + \frac{B}{2A} - \frac{\sqrt{B^2 - 4AC}}{2A} = 0$$

WHICH GIVES US:

EITHER $X = -\frac{B}{2A} + \frac{\sqrt{B^2 - 4AC}}{2A}$

OR $X = -\frac{B}{2A} - \frac{\sqrt{B^2 - 4AC}}{2A}$

ADDING

$100 + 100 + 100 = \square$

$99 + \underset{\wedge}{100} + 101 = \square$

$70 + 70 + 70 + 70 + 70 =$

$66 + 68 + \underset{\wedge}{70} + 72 + 74 =$

$196 + 197 + 198 + 199 + \underset{\wedge}{200} + 201 + 202 + 203 + 204 = \square$

$8 \quad 8\frac{1}{2} \quad \underset{\wedge}{9} \quad 9\frac{1}{2} \quad 10 =$

$55 + 57 + 59 \underset{\wedge}{+} 61 + 63 + 65 =$

$11 + 12 + 13 + 14 + \quad \wedge \quad + 28 + 29 = \square$

HINT: WHAT NUMBER IS ABOVE THE $\wedge$?

$\overbrace{15 + 19 + 23 + 27 + \cdots \wedge \cdots + 95}^{80} = \square$

(15 to 19: 4)

HOW DID WE GET 55?

HINT: HOW MANY NUMBERS ARE TO BE ADDED? $\square$

TO FIND OUT, WE LOOK AT 4 AND AT 80.

DO YOU SEE WHY THE ANSWER IS 1155?

$7 + 12 + 17 + \cdots \quad \cdots + 357 = \square$

$6 + 11 + 16 + \cdots \quad \cdots 626 = \square$

$20 + 23 + 26 + \cdots \quad \cdots + 560 = \square$

$5 + 7\frac{1}{2} + 10 + 12\frac{1}{2} + \cdots \quad + 80 = \square$

MORE ADDING

1) $3 + 4 + 5 + 6 =$

2) $11 + 12 + 13 + 14 + 15 + 16 + 17 + 18 + 19 =$

3) $2 + 6 + 10 + 14 + 18 + 22 + 26 + 30 + 34 + 38 =$

4) $3 + 6 + 9 + 12 + 15 + 18 + 21 + 24 + 27 + 30 + 33 =$

5) $23 + 25 + 27 + \cdots\cdots + 79 =$

THIS ONE IS A LITTLE HARDER

$79 = 23 + ?$ 56

SO 23 + LOTS OF 2'S = 79

THE AVERAGE TERM IS HALFWAY FROM 23 TO 79. (51)

HOW MANY TERMS ARE THERE?

HOW DO WE FIND THE SUM? DID YOU GET 1479?

6) WHAT IS THE SUM OF THE FIRST FIFTY ODD NUMBERS?

7) $\frac{1}{5} + \frac{2}{5} + \frac{3}{5} + \cdots\cdots + \frac{33}{5} =$

8) $72 + 75 + 78 + \cdots + 408 =$

9) WHAT IS WRONG WITH THIS PROBLEM?

$14 + 16 + 18 + 20 + \cdots\cdots + 315 =$

10) $\frac{1}{15} + \frac{2}{15} + \frac{4}{15} + \frac{8}{15} + \cdots\cdots + \frac{2048}{15} =$

SOME ADDING VARIETY

1) $1+2+3+4+5+6+\cdots+50=$

2) $70+80+90+\cdots+570=$

3) $18+23+28+33+\cdots+408=$

4) $3+6+9+\cdots\cdots+99=$

(HINT: IN PROBLEMS LIKE THESE, IT SOMETIMES HELPS TO WRITE THE SAME SERIES IN REVERSE AND ADD THE TWO SERIES:

$99+96+93+90+\cdots 9+6+3$

5) $\frac{3}{4}+1+1\frac{1}{4}+1\frac{1}{2}+1\frac{3}{4}+\cdots+21\frac{1}{2}=$

6) $4+4\frac{1}{2}+5+5\frac{1}{2}+6+\cdots+52\frac{1}{2}=$

7) $11+12\frac{1}{3}+13\frac{2}{3}+15+16\frac{1}{3}+\cdots+39=$

. . . .

8) $2+4+8+16+32+\cdots+32768=$

9) $1-\frac{1}{2}+\frac{1}{4}-\frac{1}{8}+\frac{1}{16}-\frac{1}{32}+\cdots\cdots-\frac{1}{2048}=$

10) $1-\frac{1}{2}+\frac{1}{4}-\frac{1}{8}+\frac{1}{16}-\frac{1}{32}+\cdots\cdots-\frac{1}{2048}+\cdots=$

WHY IS THERE A SPACE BETWEEN # 7 AND # 8?

IN PROBLEM 10, WHAT DOES THAT FINAL PLUS SIGN DO?

ETH OZO

XO
OX

WLO
LWO
OLW
OWL
LOW
WOL

LWFO

(PERHAPS OUR FRIEND THE OWL CAN BE OF SOME ASSISTANCE HERE!)

EBRZA
ARBEZ
ABZER
BAREZ
ZBREA
RAZBE
AZBRE

HOW MANY PERMUTATIONS OF FOWL BEGIN WITH F?

NOW, HOW MANY PERMUTATIONS OF FOWL BEGIN WITH O?
W?
L?

SO, HOW MANY PERMUTATIONS HAS FOWL?

HOW MANY PERMUTATIONS HAS LION?
(DOES THE FACT THAT A LION IS USUALLY BIGGER THAN A FOWL AND MIGHT EAT IT AFFECT THE ANSWER?)

HOW MANY PERMUTATIONS HAS ZEBRA?
(HINT: HOW MANY OF THEM BEGIN WITH Z?)
AND THIS MEANS: HOW MANY PERMUTATIONS HAS BEAR?)

JUST AS OWL HELPS US TO FIND FOWL,
MAYBE BEAR CAN HELP US TO FIND ZEBRA.

SOMETIMES IT HELPS TO MAKE A CHART:

XO	2
WOL	6
OLNI	
BRAZE	
URSLAW	
RADPOLE	
GOLFMAIN	

HERE ARE SOME PRACTICE PROBLEMS:

AFEL BACOR RLDAIZ
MLACE HOTNYP SHORTIC
ONDERFUL EMPIGA

HOW MANY PERMUTATIONS HAS GNU

GNU
UNG
NGU
UGN
GUN
NUG

THE GNU HAS
6 DIFFERENT WAYS
TO PERMUTE ITS LETTERS.

WHAT ABOUT OX

OX XO

WHAT ABOUT THE BUG

IS THERE SOME REASON
WHY BUG HAS ONLY
5 PERMUTATIONS?
DID WE MISS ONE?

GBU
UBG
GUB
BGU
UGB

WORM

HOW MANY PERMUTATIONS
OF WORM
BEGIN WITH W?

WORM
WOMR
WROM
WRMO
WMOR
WMRO

WHAT ABOUT LOON?

DOES LOON HAVE THE SAME NUMBER OF PERMUTATIONS AS LION? CAN YOU WRITE ALL 24?

TO ANSWER THIS, WE PUT A TAIL INTO ONE OF THE O'S OF LOON: THIS GIVES US THE LOQN.

LOQN HAS 24 PERMUTATIONS. WE LIST THEM, PUTTING TOGETHER THE WORDS WITH O AND Q EXCHANGED:

LOQN ← WE GET THIS LIST
LQON
NOLQ
NQLO
OQLN
QOLN
NLOQ
NLQO
OLQN
QLON
⋮

WE THEN REMOVE THE TAILS FROM ALL THE Q'S TO GET THIS LIST →

LOON
LOON
NOLO
NOLO
OOLN
OOLN
NLOO
NLOO
OLON
OLON
⋮

FROM THIS PROCEDURE IT IS CLEAR THAT THE NUMBER OF PERMUTATIONS OF LOON IS 12.

THERE ARE, OF COURSE, LARGER ANIMALS

HOW MANY PERMUTATIONS HAS

DODO ? ANACONDA ? OPOSSUM?
TREETOAD ? BEETLE ? ONDERFUL?
DIKDIK ELEPHANT ?

FROM A DECK OF CARDS, TAKE ALL 13 OF THE HEARTS.

HOW MANY DIFFERENT SETS OF 5 HEARTS COULD HAVE BEEN SELECTED?

TRY LOOKING AT THE PROBLEM THIS WAY: THE A 7 10 J K CAN BE REPRESENTED BY THE FOLLOWING DIAGRAM:

A♡	2♡	3♡	4♡	5♡	6♡	7♡	8♡	9♡	10♡	J♡	Q♡	K♡
YES	NO	NO	NO	NO	NO	YES	NO	NO	YES	YES	NO	YES

THIS HAND, THEREFORE, IS SEEN SIMPLY AS
Y N N N N N Y N N Y Y N Y

OR, MORE SIMPLY: YNNNNNYNNYYNY

SO THE QUESTION BECOMES:
"HOW MANY 13-LETTER WORDS ARE THERE SPELT WITH 8 N'S AND 5 Y'S?"

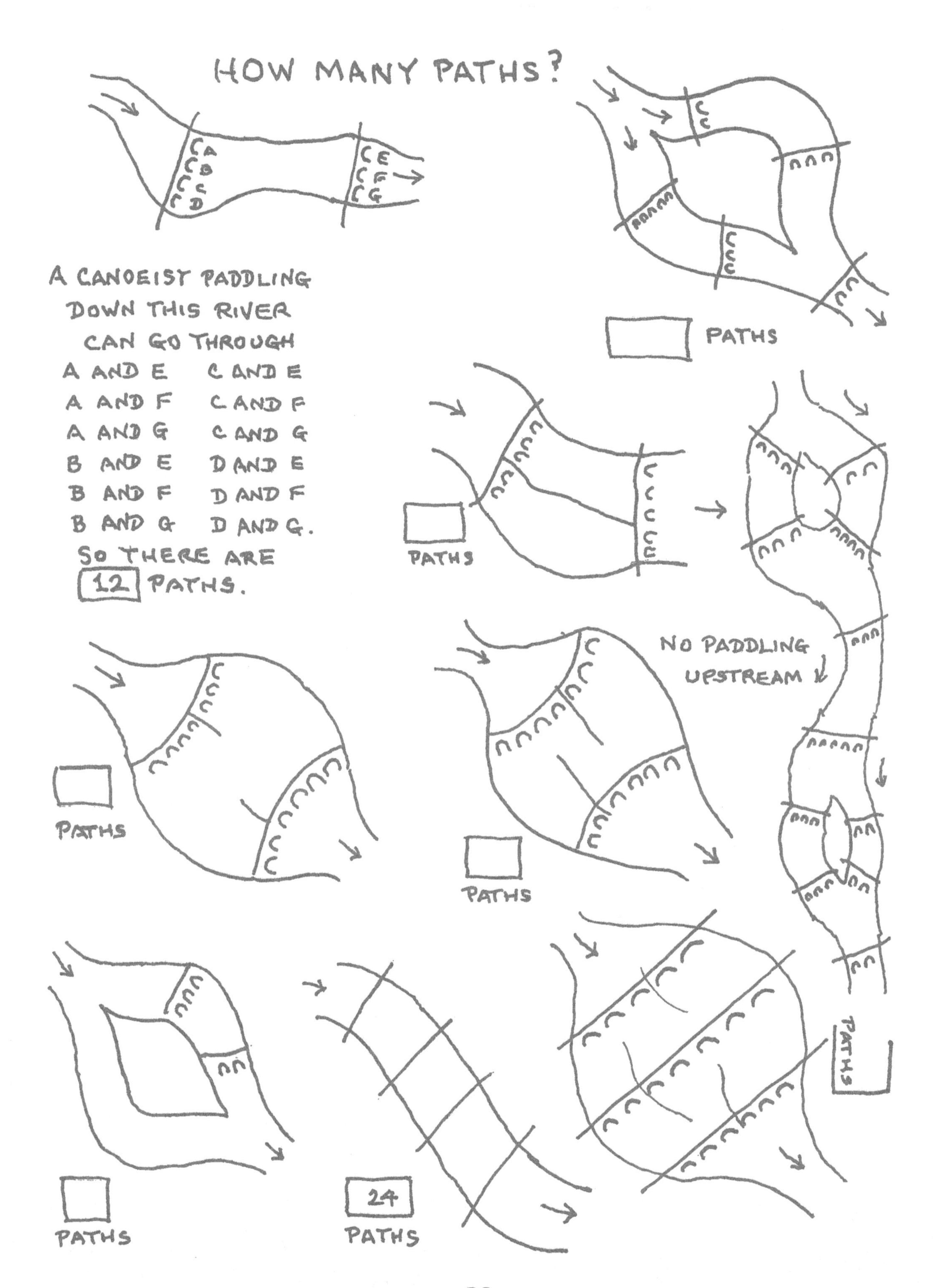
HOW MANY PATHS?
A
B
C
D
E
F
G
A CANOEIST PADDLING DOWN THIS RIVER CAN GO THROUGH
A AND E C AND E
A AND F C AND F
A AND G C AND G
B AND E D AND E
B AND F D AND F
B AND G D AND G.
SO THERE ARE
12
PATHS.
PATHS
PATHS
NO PADDLING UPSTREAM
PATHS
PATHS
PATHS
PATHS
24
PATHS

IF WE THROW ONE DIE

THERE ARE 6 POSSIBLE RESULTS.

ASSUMING THAT EACH FACE
IS EQUALLY LIKELY
TO APPEAR ON TOP,

PROBABILITIES ARE

$1/6$

$1/6$

$1/6$

$1/6$

$1/6$

$1/6$

DIE A

1 2 3 4 5 6

IF WE THROW
DIE A TWICE
AND ADD
THE TWO FACES:

p(2) =
p(3) =
p(4) =
p(5) =
p(6) =
p(7) =
p(8) =
p(9) =
p(10) =
p(11) =
p(12) =

DIE B

1 2 3 4 5 6

DIE A DIE B

WE HAVE 2 DICE TO THROW:
ONE WITH 6 SIDES
(SIDES EQUALLY LIKELY)
OTHER WITH 7 SIDES
(SIDES EQUALLY LIKELY)
AND ADD THE TWO FACES:

p(2) =
p(3) =
p(4) =
⋮
p( ) =

IF WE THROW
DIE A THREE TIMES
AND ADD
THE THREE FACES

p(3)
p(4)
p(5)
p(6)
⋮
p(18)

PRIME FACTORS

IF A WHOLE NUMBER, (AN INTEGER)
CAN BE FOUND BY MULTIPLYING TWO INTEGERS,
EACH GREATER THAN 1,
IT IS CALLED A <u>COMPOSITE</u> NUMBER.

A WHOLE NUMBER WHICH IS GREATER THAN 1,
BUT NOT COMPOSITE, IS CALLED A <u>PRIME</u> NUMBER.

(JUST AS ADDING ZERO TO A NUMBER
DOES NOT CHANGE THE NUMBER,
MULTIPLYING A NUMBER BY 1
DOES NOT CHANGE THE NUMBER.)

2 IS PRIME
3 IS PRIME
4
5 IS PRIME
6
7 IS PRIME
8
9
10
11 IS PRIME
12
13 IS PRIME
14
15
16
17 IS PRIME
18
19 IS PRIME
20
21
22
23 IS PRIME
24
25

QUESTION:

WHICH OF THE FOLLOWING NUMBERS ARE PRIME?

110 111 112

113 114 115

116 117 118

119 120

PRIME FACTORS

3 TIMES 5 = 15   11 TIMES 2 = 22   4 TIMES 8 =

10 TIMES 6 =   9 TIMES 4 =   6 TIMES 13 =

SOMETIMES IT'S MORE INTERESTING TO GIVE THE ANSWER

AND ASK FOR THE QUESTION.

TIMES = 21

OF COURSE 1 TIMES 21 WILL DO
BUT IS UNINTERESTING.
WHAT WE WANT IS 3 TIMES 7.

TIMES = 55

IS THERE SOMETIMES A CHOICE?

TAKE: TIMES = 30

WE CAN WRITE 2 TIMES 15 = 30
3 TIMES 10 = 30
OR 5 TIMES 6 = 30

SO THERE SEEMS TO BE A CHOICE.

BUT WE CONTINUE:

2 TIMES (3 TIMES 5) = 30
3 TIMES (2 TIMES 5) = 30
OR 5 TIMES (2 TIMES 3) = 30

SO, NO MATTER HOW WE START,
WE ARRIVE AT THE SAME THREE FACTORS.
ARRANGING THEM IN ORDER OF SIZE:
2 TIMES 3 TIMES 5 = 30.

HERE ARE SOME OTHERS:

42	42	42	100	100
(6)(7)	(3)(14)	(2)(21)	(10)(10)	(4)(25)
(2)(3)(7)	(3)(2)(7)	(2)(3)(7)	(2)(5)(2)(5)	(2)(2)(5)(5)

140 81 225 72 128 2310

HOW MANY DIVISORS HAS 675?

1, 3, 5, 9, 15, 675, ... OOPS, I THINK WE'VE MISSED SOME.

WHAT ARE THE FACTORS OF 675?

675
(3)(225)
(3)(3)(75)
(3)(3)(3)(25)
(3)(3)(3)(5)(5)

SO ANY DIVISOR OF 675 MUST HAVE, AS FACTORS: 3'S AND/OR 5'S.

THE NUMBER OF 3'S MUST BE 0, 1, 2, OR 3 AND THE NUMBER OF 5'S MUST BE 0, 1, OR 2 SO WE MAKE AN ARCH DIAGRAM:

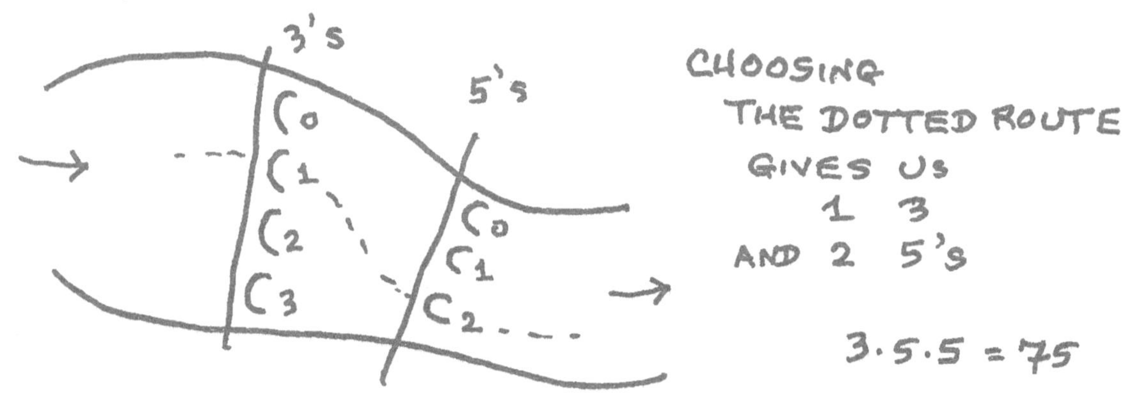

CHOOSING THE DOTTED ROUTE GIVES US 1 3 AND 2 5'S

3·5·5 = 75

HOW MANY ROUTES ARE THERE? (DIVISORS OF 675)

NOTE: EVERY NUMBER HAS TWO UNINTERESTING DIVISORS: ITSELF AND 1. WE DON'T LIST THEM. A NUMBER WITH NO DIVISORS IS SAID TO BE PRIME.

WE HAVE, ON THE OTHER PAGE,
FACTORED 675 COMPLETELY
INTO ITS PRIME FACTORS.

CAN YOU DO THE SAME FOR:

1) 21
2) 70
3) 28
4) 44
5) 144
6) 130
7) 210
8) 196
9) 6050
10) 3528

THE NUMBER 24 HAS 6 DIVISORS.
THEY ARE: 2, 3, 4, 6, 8, 12

HOW MANY DIVISORS HAVE

1) 27
2) 36
3) 75
4) 143
5) 144
6) 196
7) 256
8) 40320
9) 65536
10) 72000

WHAT NUMBER LESS THAN A MILLION HAS THE MOST DIVISORS?

WHAT NUMBER LESS THAN A MILLION
HAS THE MOST DIFFERENT DIVISORS?

```
x x x x x
x x x x x
x x x x x
x x x x x
x x x x x
```

THE DRAWING AT LEFT REPRESENTS
5 TIMES 5.

THE DRAWING AT RIGHT
REPRESENTS 6 TIMES 4.

```
o o o o o o
o o o o o o
o o o o o o
o o o o o o
```

IF WE PLACE ONE DRAWING ON TOP OF THE OTHER,

```
⊗ ⊗ ⊗ ⊗ ⊗ o
⊗ ⊗ ⊗ ⊗ ⊗ o
⊗ ⊗ ⊗ ⊗ ⊗ o
⊗ ⊗ ⊗ ⊗ ⊗ o
x x x x x
```

IT IS EASY TO SEE WITHOUT COUNTING
THAT THE NUMBER OF O'S
IS 1 LESS
THAN THE NUMBER OF X'S

SO (5 TIMES 5) − 1 = (6 TIMES 4).

WE NOW COMPARE 5 TIMES 5 WITH 7 TIMES 3

```
x x x x x
x x x x x
x x x x x
x x x x x
x x x x x
```

5 TIMES 5

```
o o o o o o o
o o o o o o o
o o o o o o o
```

```
⊗ ⊗ ⊗ ⊗ ⊗ o o
⊗ ⊗ ⊗ ⊗ ⊗ o o
⊗ ⊗ ⊗ ⊗ ⊗ o o
x x x x x
x x x x x
```

WE SEE
THAT THE NUMBER OF O'S
IS 4 LESS
THAN THE NUMBER OF X'S

SO (5 TIMES 5) − 4 = 7 TIMES 3.

HOW DOES 5 TIMES 5 COMPARE WITH 8 TIMES 2?
USING X'S AND O'S AS ABOVE, CAN YOU DRAW THE DIAGRAM?

SO (5 TIMES 5) − [   ] = 8 TIMES 2

10 TIMES 10 = ☐	20 TIMES 20 = ☐	25 × 25 = ☐
11 TIMES 9 = ☐	21 TIMES 19 = ☐	TIMES = ☐
12 TIMES 8 = ☐	= ☐	TIMES = ☐

USE SIMILAR DIAGRAMS TO SHOW THAT:
5 TIMES 5 IS 9 MORE THAN 8 TIMES 2
5 TIMES 5 IS 16 MORE THAN 9 TIMES 1

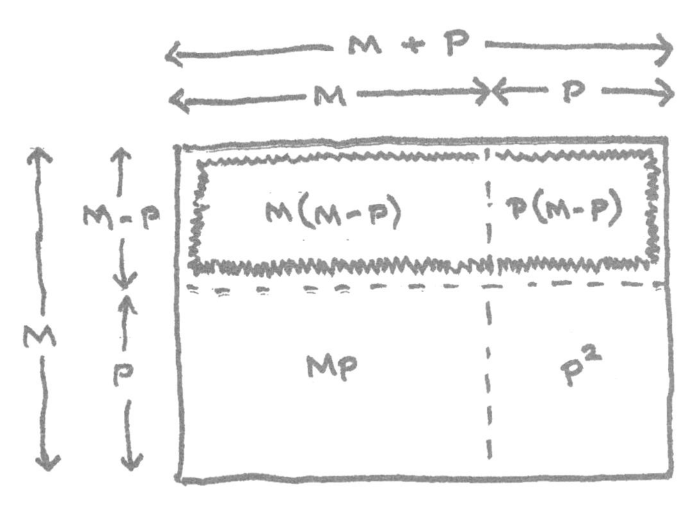

$M(M-P) + P(M-P)$
$= (M+P)(M-P)$

MULTIPLY THE FOLLOWING:

1) 28 TIMES 32
2) 14 TIMES 26
3) 6 TIMES 14
4) 43 TIMES 37
5) 49 TIMES 51
6) 67 TIMES 72

WHY IS PROBLEM 6 DIFFERENT FROM THE OTHERS?

IF WE HAVE Q+7 OF ANY ITEM,
WE HAVE Q OF THEM PLUS 7 OF THEM.
IF WE HAVE Q+7 OF ITEM K,
WE HAVE Q K'S PLUS 7 K'S
SO IF WE HAVE Q+7 OF ITEM Q-7
THAT MEANS WE Q(Q-7) + 7(Q-7)
(WE HAVE DISTRIBUTED THE Q+7).

$$\begin{aligned}(Q+7)(Q-7) &= Q(Q-7) + 7(Q-7) \\ &= Q^2 - 7Q + 7Q - 49 \\ &= Q^2 - 49\end{aligned}$$

MULTIPLY THE FOLLOWING:

1) 28 TIMES 32
2) 14 TIMES 26
3) 6 TIMES 14
4) 43 TIMES 37
5) 49 TIMES 51
6) 67 TIMES 72

TRY THESE:

7) (8+5)(8-5)
8) (Z+2)(Z-2)
9) (M-3)(M+3)
10) (W-6)(W+4)
11) (@+3)(@-5)
12) (2△+3)(5△+7)
13) (*-8)(*+11)
14) (↓+4)(↓-9)
15) (□+7)□

CHAPTER 1

7 TIMES 7 IS
8 TIMES 6 IS

10 TIMES 10 IS
11 TIMES 9 IS

5 TIMES 5 IS
6 TIMES 4 IS

12 TIMES 12 IS
13 TIMES 11 IS

20 TIMES 20 IS
21 TIMES 19 IS

50 TIMES 50 IS
TIMES IS

CHAPTER 2

4 TIMES 4 =
6 TIMES 2 =

8 TIMES 8 =
10 TIMES 6 =

9 TIMES 9 =
11 TIMES 7 =

7 TIMES 7 =
9 TIMES 5 =

6 TIMES 6 =

TIMES =
42 TIMES 38 =

CHAPTER 3

5 TIMES 5 =
9 TIMES 2 =

8 TIMES 8 =
11 TIMES 5 =

10 TIMES 10 =
13 TIMES 7 =

20 TIMES 20 =

16 TIMES 10 =

73 TIMES 67 =

CHAPTER 4

30 TIMES 30 =
34 TIMES 26 =

J TIMES J =
J+4 TIMES J-4 =

• • • • •

CHAPTER 5

TIMES =

CHAPTER 6

TIMES =

CHAPTER W

11 TIMES 11 =
(11+W)(11-W) =

CHAPTER Q

P TIMES P =
(P+Q) TIMES (P-Q) =

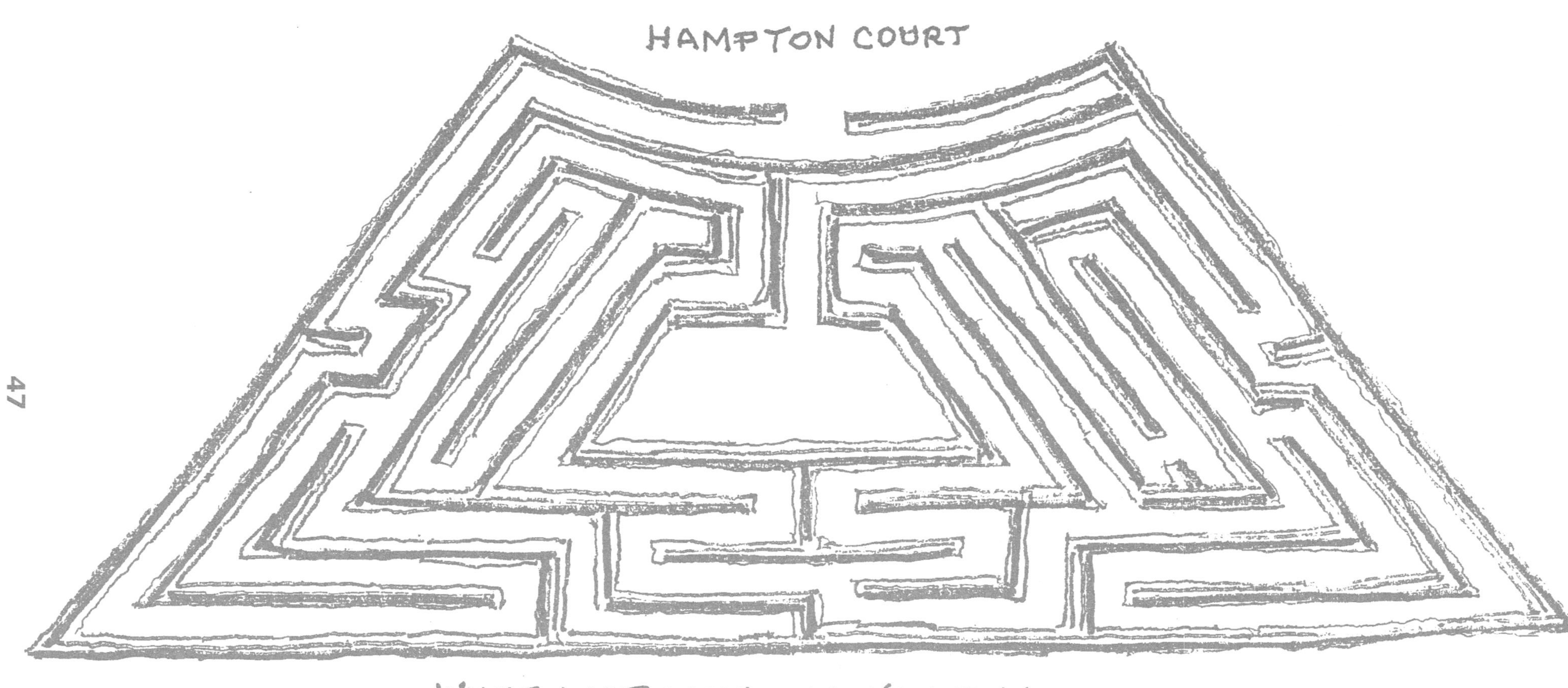

WHAT INSTRUCTIONS WOULD YOU GIVE
TO A VISITOR AT THE GATE
FOR HOW TO REACH THE CENTRE WITHOUT HAVING TO RETRACE?

HINT: THIS MAZE HAS NO 4-WAY INTERSECTIONS.

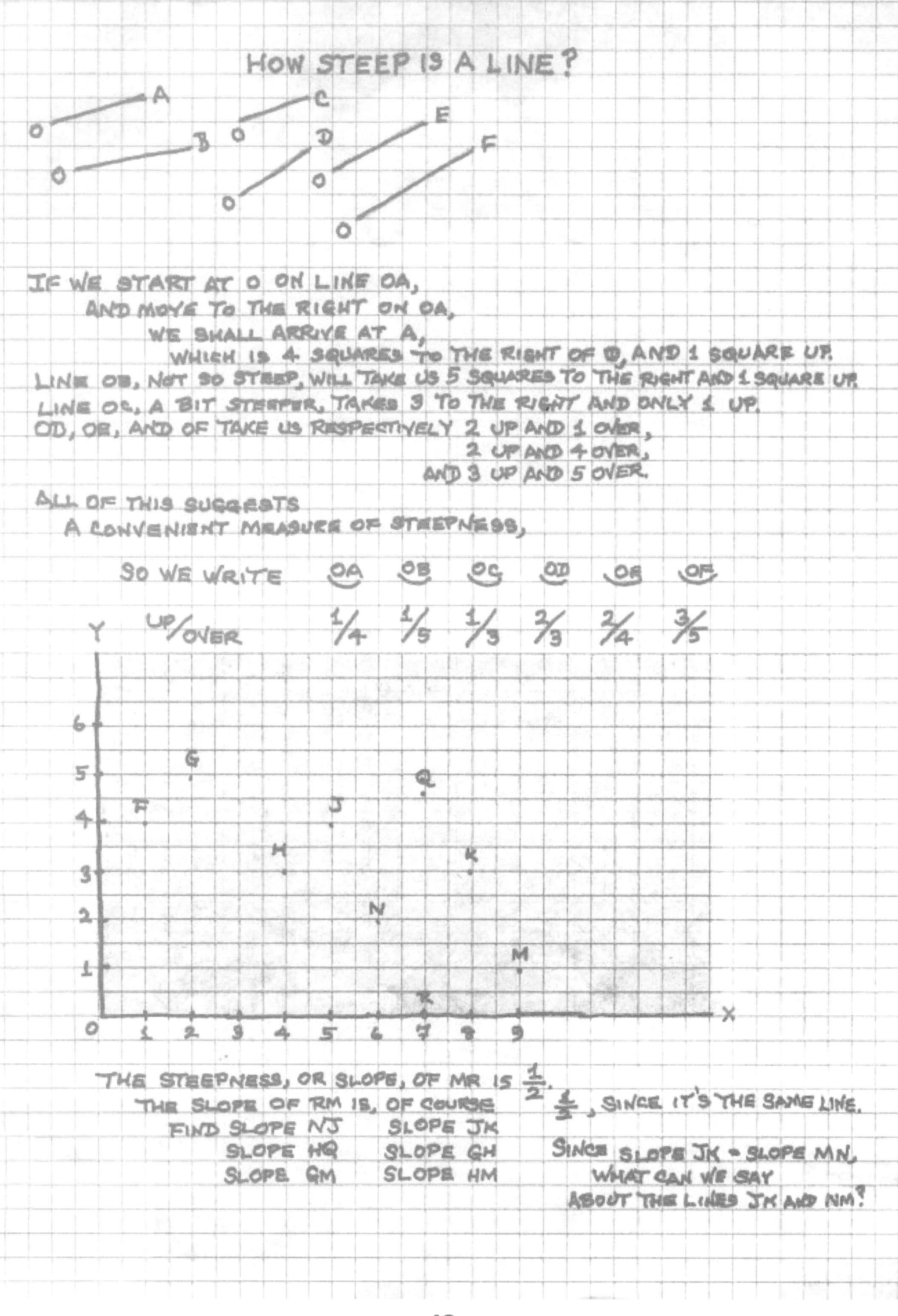

HOW STEEP IS A LINE?

IF WE START AT O ON LINE OA,
AND MOVE TO THE RIGHT ON OA,
WE SHALL ARRIVE AT A,
WHICH IS 4 SQUARES TO THE RIGHT OF O, AND 1 SQUARE UP.
LINE OB, NOT SO STEEP, WILL TAKE US 5 SQUARES TO THE RIGHT AND 1 SQUARE UP.
LINE OC, A BIT STEEPER, TAKES 3 TO THE RIGHT AND ONLY 1 UP.
OD, OE, AND OF TAKE US RESPECTIVELY 2 UP AND 1 OVER,
2 UP AND 4 OVER,
AND 3 UP AND 5 OVER.

ALL OF THIS SUGGESTS
A CONVENIENT MEASURE OF STEEPNESS,

SO WE WRITE	OA	OB	OC	OD	OE	OF
UP/OVER	1/4	1/5	1/3	2/3	2/4	3/5

THE STEEPNESS, OR SLOPE, OF MR IS $\frac{1}{2}$.
THE SLOPE OF RM IS, OF COURSE $\frac{1}{2}$, SINCE IT'S THE SAME LINE.

FIND SLOPE NJ   SLOPE JK
SLOPE HQ   SLOPE GH
SLOPE GM   SLOPE HM

SINCE SLOPE JK = SLOPE MN,
WHAT CAN WE SAY
ABOUT THE LINES JK AND NM?

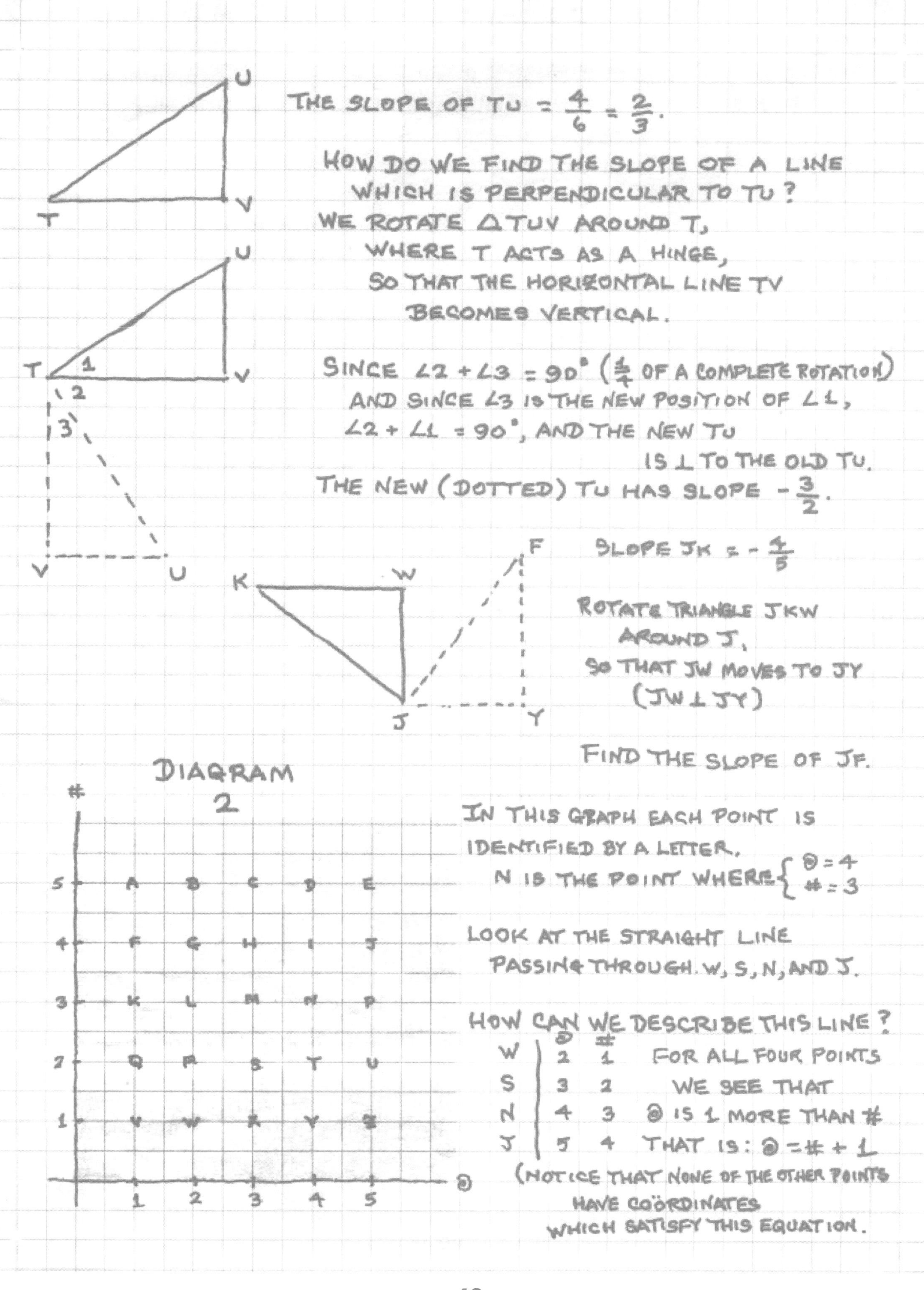

THE SLOPE OF TU $= \frac{4}{6} = \frac{2}{3}$.

HOW DO WE FIND THE SLOPE OF A LINE WHICH IS PERPENDICULAR TO TU?

WE ROTATE $\triangle$TUV AROUND T, WHERE T ACTS AS A HINGE, SO THAT THE HORIZONTAL LINE TV BECOMES VERTICAL.

SINCE $\angle 2 + \angle 3 = 90°$ ($\frac{1}{4}$ OF A COMPLETE ROTATION) AND SINCE $\angle 3$ IS THE NEW POSITION OF $\angle 1$, $\angle 2 + \angle 1 = 90°$, AND THE NEW TU IS $\perp$ TO THE OLD TU.

THE NEW (DOTTED) TU HAS SLOPE $-\frac{3}{2}$.

SLOPE JK $= -\frac{4}{5}$

ROTATE TRIANGLE JKW AROUND J, SO THAT JW MOVES TO JY (JW $\perp$ JY)

FIND THE SLOPE OF JF.

DIAGRAM 2

IN THIS GRAPH EACH POINT IS IDENTIFIED BY A LETTER.
N IS THE POINT WHERE $\begin{cases} @ = 4 \\ \# = 3 \end{cases}$

LOOK AT THE STRAIGHT LINE PASSING THROUGH W, S, N, AND J.

HOW CAN WE DESCRIBE THIS LINE?

	@	#
W	2	1
S	3	2
N	4	3
J	5	4

FOR ALL FOUR POINTS WE SEE THAT @ IS 1 MORE THAN #
THAT IS: @ = # + 1

(NOTICE THAT NONE OF THE OTHER POINTS HAVE COÖRDINATES WHICH SATISFY THIS EQUATION.

1) WHAT IS THE SLOPE OF THE LINE EA?

2) WHAT IS THE SLOPE OF AB?
(SAME ANSWER, SINCE IT'S THE SAME LINE.)

3) WHAT IS THE SLOPE OF MD?

4) WHAT IS THE SLOPE OF MH?
(NO ANSWER: VERTICAL LINES
DON'T HAVE SLOPES.)

5) BESIDES KHE, WHAT LINES HAVE SLOPE $\frac{1}{2}$?

6) WHAT LINES HAVE SLOPE $-\frac{2}{3}$?

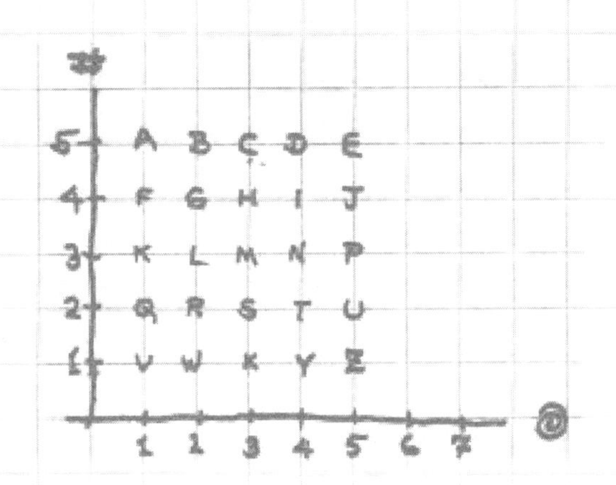

CAN YOU FIND THE EQUATION OF THE LINE CONTAINING

[ ] K, R, AND X ? SLOPE = [ ]

[ ] Q, L, H, AND D ? SLOPE = [ ]

[ ] X, N, AND E ? SLOPE = [ ]

[ ] Q, M, AND J ? SLOPE = [ ]

- 3 = @ [ ] SLOPE = [ ]

2# = @ [ ] SLOPE = [ ]

[ ] [K,] SLOPE = $\frac{2}{3}$

2@ = # + 1 [ ] SLOPE = [ ]

WHAT ARE THE COÖRDINATES OF C ?

THEY ARE: @ = 3
= 5

SO, THE COORDINATES OF C ARE {3, 5}

(WE ALWAYS SAY THE HORIZONTAL COÖRDINATE FIRST.)

SOME GEOMETRY

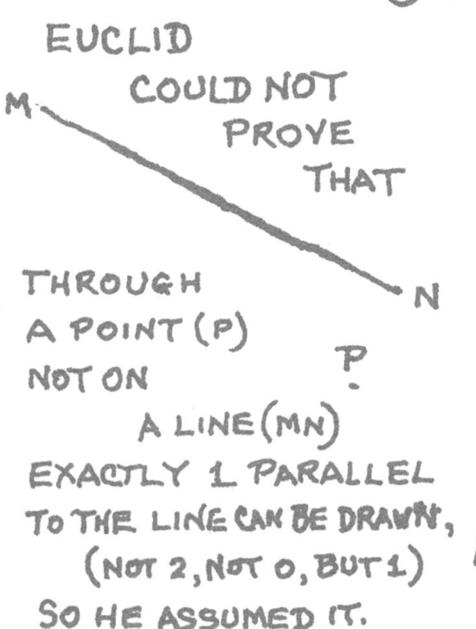

EUCLID
COULD NOT
PROVE
THAT

THROUGH
A POINT (P)
NOT ON
A LINE (MN)
EXACTLY 1 PARALLEL
TO THE LINE CAN BE DRAWN,
(NOT 2, NOT 0, BUT 1)
SO HE ASSUMED IT.

FROM THIS ASSUMPTION
(THE EUCLIDEAN ASSUMPTION)
IT CAN EASILY BE SHOWN THAT
IF TWO LINES ARE PARALLEL,
THEN THE Z ANGLES ARE EQUAL:
$\angle 1 = \angle 2$
AND FROM THIS
Z ANGLE THEOREM
IT FOLLOWS EASILY
THAT
THE SUM OF THE ANGLES OF A TRIANGLE
IS EQUAL TO HALF A ROTATION
(USUALLY CALLED 180 DEGREES)

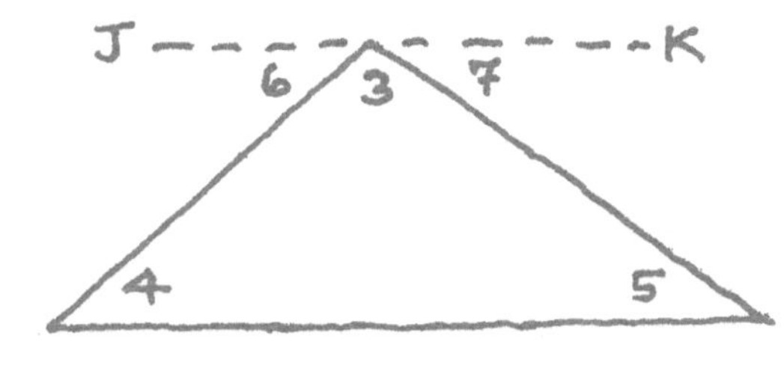

FOR WE CAN DRAW A PARALLEL JK
AND SAY $\angle 6 + \angle 3 + \angle 7 = 180°$
$\angle 6 = \angle 4 \qquad \angle 7 = \angle 5$
THEREFORE $\angle 4 + \angle 3 + \angle 5 = 180°$

IF A CHORD (AQ)
AND A DIAMETER (BOQ)
MEET ON A CIRCLE (AT Q),
IT'S EASY TO SHOW
THAT THE ANGLE AT C
(CALLED AN INSCRIBED ANGLE)
IS HALF AS BIG AS $\angle AOB$.
FOR $\angle A + \angle C + W = 180°$
AND $V + W = 180°$
SO $\angle A + \angle C + W = V + W$
AND $\angle A + \angle C = V$.
$\angle A = \angle C$ GIVES US
$\angle C + \angle C = V$

AND $\angle C = \frac{1}{2} V$
$= \frac{1}{2} \overset{\frown}{AB}$
LIKEWISE, $\angle H = \frac{1}{2} \overset{\frown}{BD}$
ADDING, WE GET
$\angle C + \angle H = \frac{1}{2}(\overset{\frown}{AB} + \overset{\frown}{BD})$
$= \frac{1}{2} \overset{\frown}{AD}$

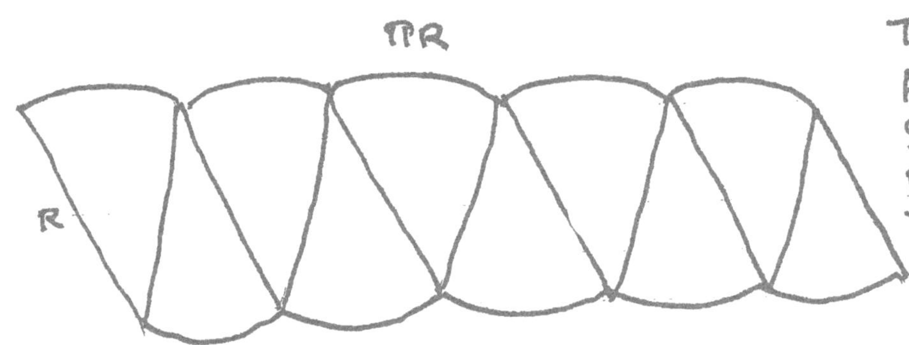

TO LEARN ABOUT THE AREA OF THE CIRCLE, WE SLICE INTO TEN EQUAL PARTS AND ARRANGE THEM AS SHOWN

THIS LOOKS SOME-WHAT LIKE A RECTANGLE

πR

R

LET'S CUT IT INTO TWICE AS MANY PARTS

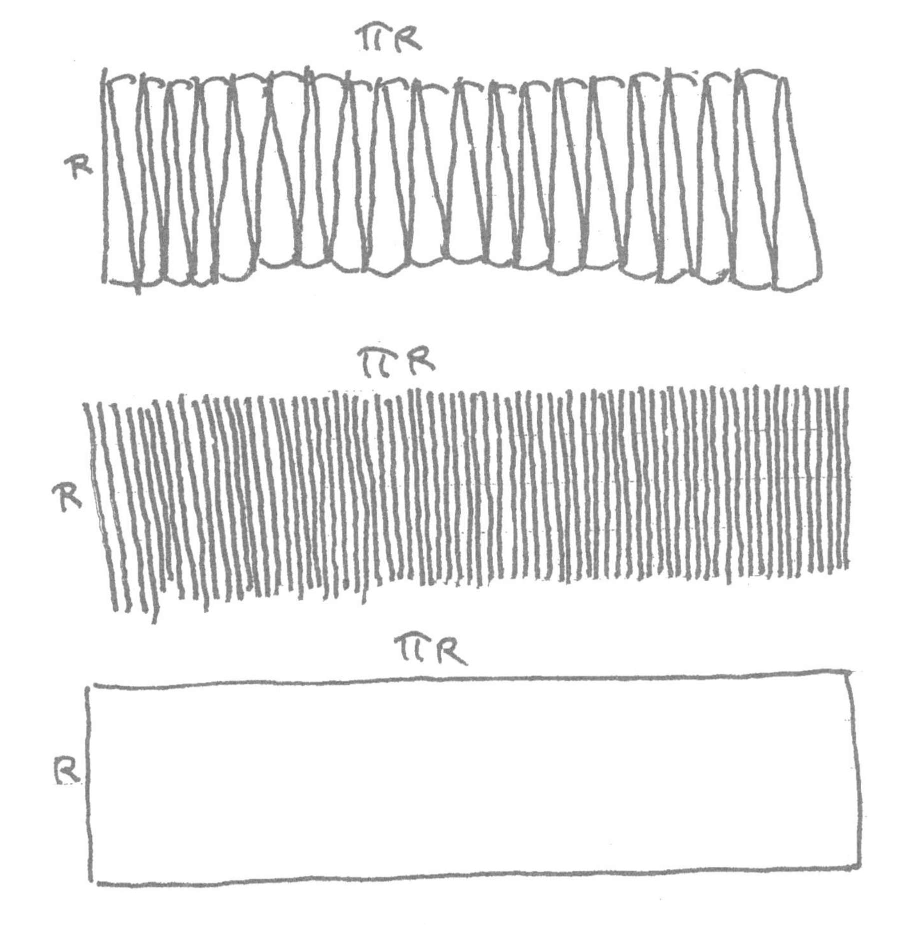

FOR CENTURIES IT HAS BEEN WELL KNOWN THAT THE LENGTH OF THE CIRCUMFERENCE IS MORE THAN THREE TIMES THE DIAMETER.

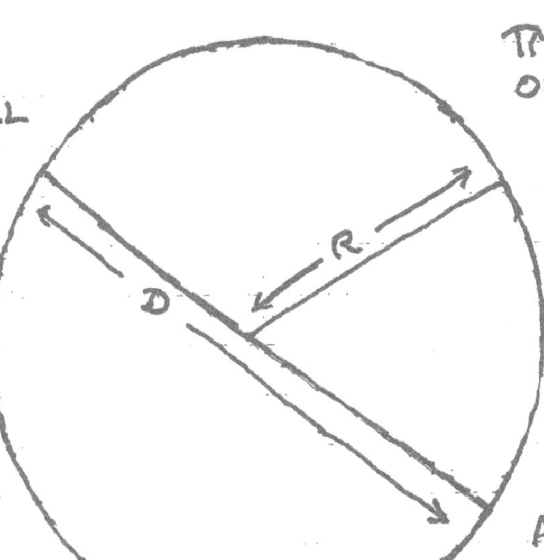

π IS NOT A FRACTION OR A REPEATING DECIMAL.

THE ANCIENT GREEKS CALCULATED π AS

$$\frac{22}{7} = 3.\overline{142857}.$$

THE EXACT NUMBER BY WHICH THE DIAMETER MUST BE MULTIPLIED TO PRODUCE THE CIRCUMFERENCE IS CALLED "PI" PRONOUNCED PYE, A LETTER FROM THE GREEK ALPHABET. IT IS WRITTEN π.

A FRACTION WHICH IS EVEN CLOSER CAN BE FOUND BY TAKING PAIRS OF THE FIRST THREE ODD NUMBERS AND WRITING THEM THUS:

$$\frac{355}{113} = 3.14159292\ldots$$

$$\pi = 3.14159265358979323846264 33\ldots$$

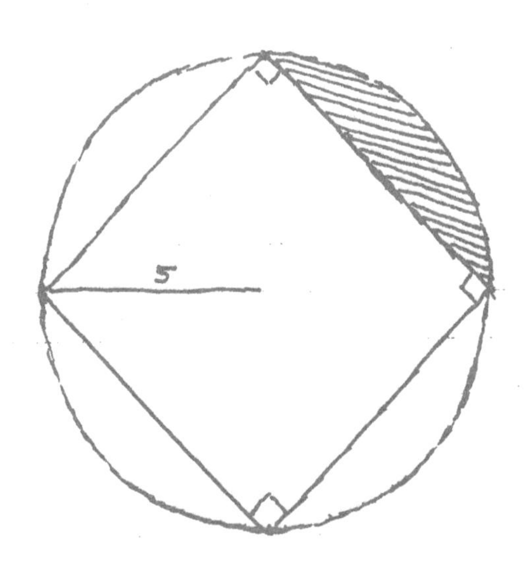

FIND SHADED AREA.

FIND SHADED AREA.

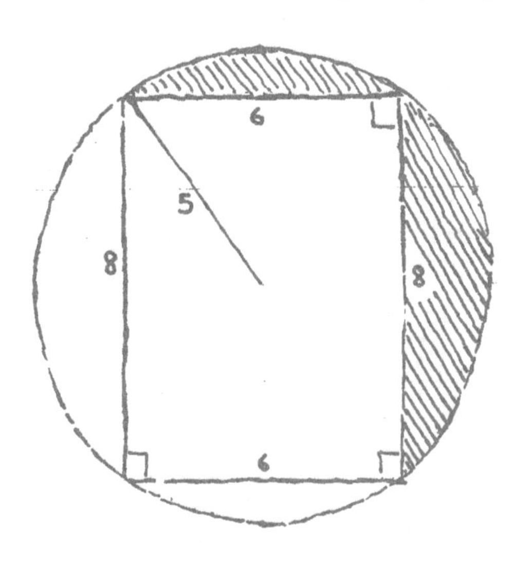

IF R = 6, ABOVE;

D =

C =

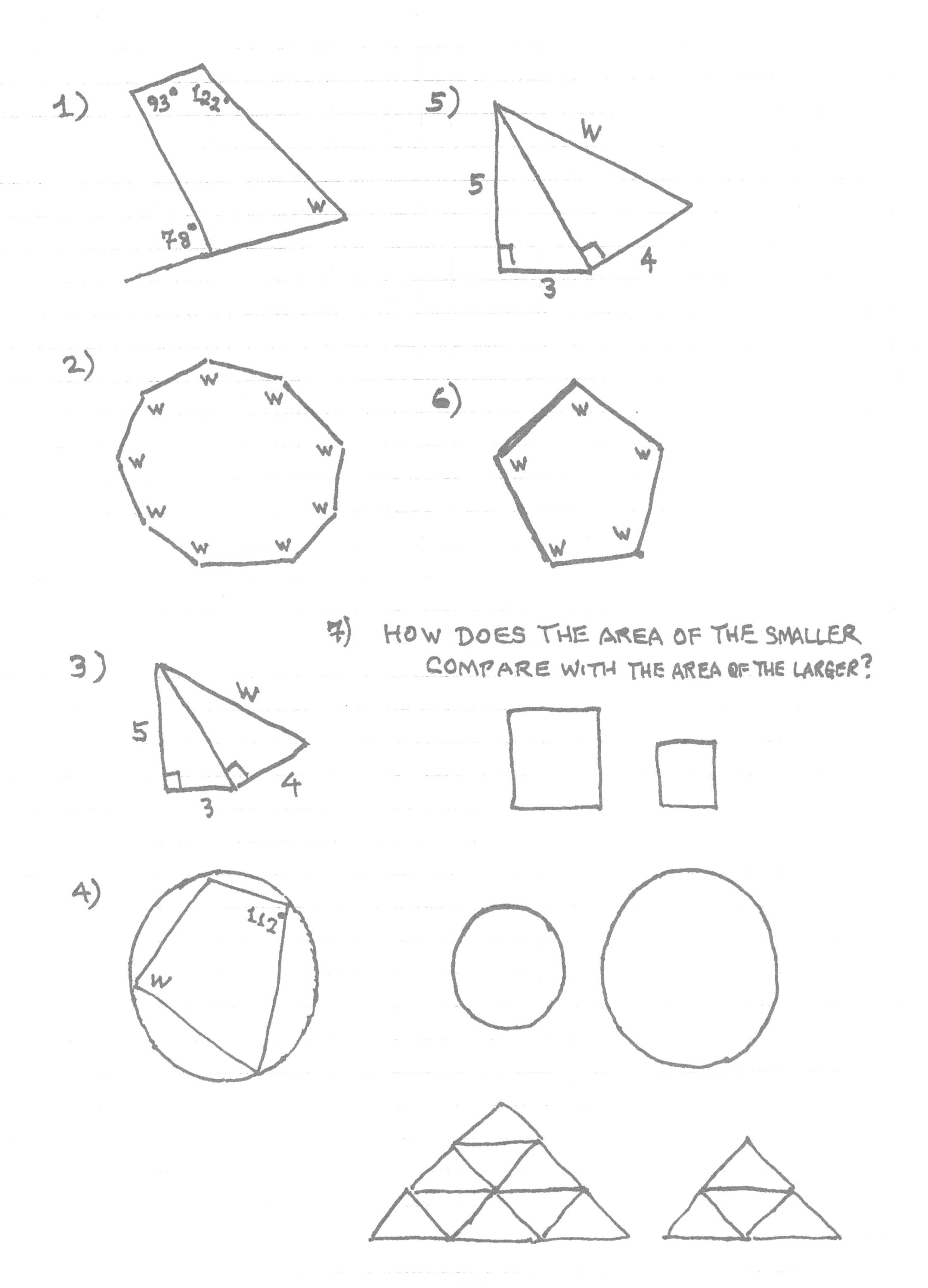
1)
93°
122°
W
78°
5)
W
5
4
3
2)
W
6)
W
3)
W
5
3
4
4)
112°
W
7) HOW DOES THE AREA OF THE SMALLER COMPARE WITH THE AREA OF THE LARGER?

TWO TRIANGLES ARE SAID TO BE SIMILAR

IF ONE IS A PHOTOGRAPHIC ENLARGEMENT OF THE OTHER.

IN OTHER WORDS, THE THREE ANGLES OF ONE MUST BE THE SAME, RESPECTIVELY, AS THE OTHER.

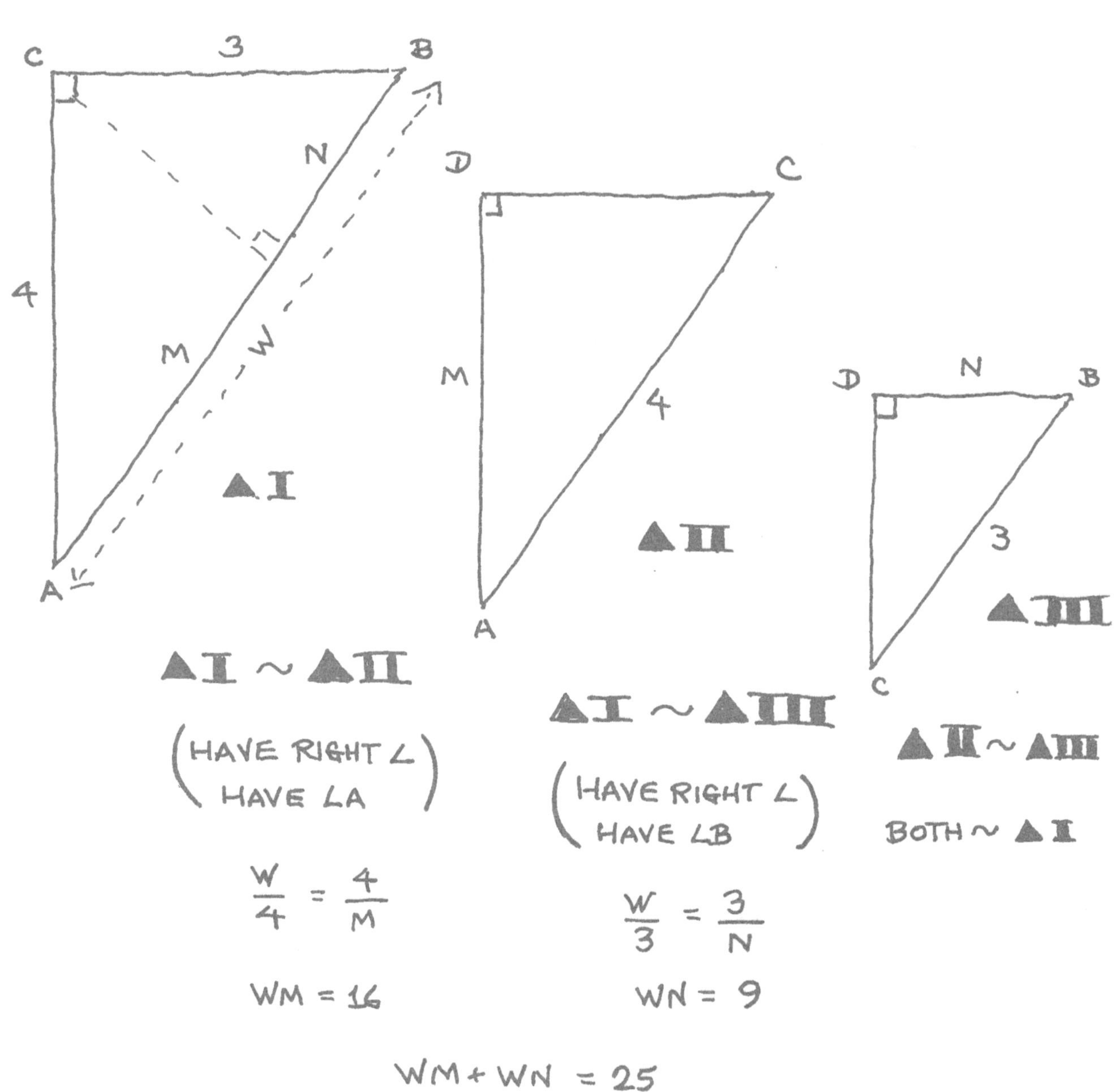

$$W(M+N) = 25$$

$$W^2 = 25$$

$$\boxed{W = 5}$$

PYTHAGORAS

AREAS

THE NUMBER OF UNITS OF AREA IN THE RECTANGLE BELOW IS 6.

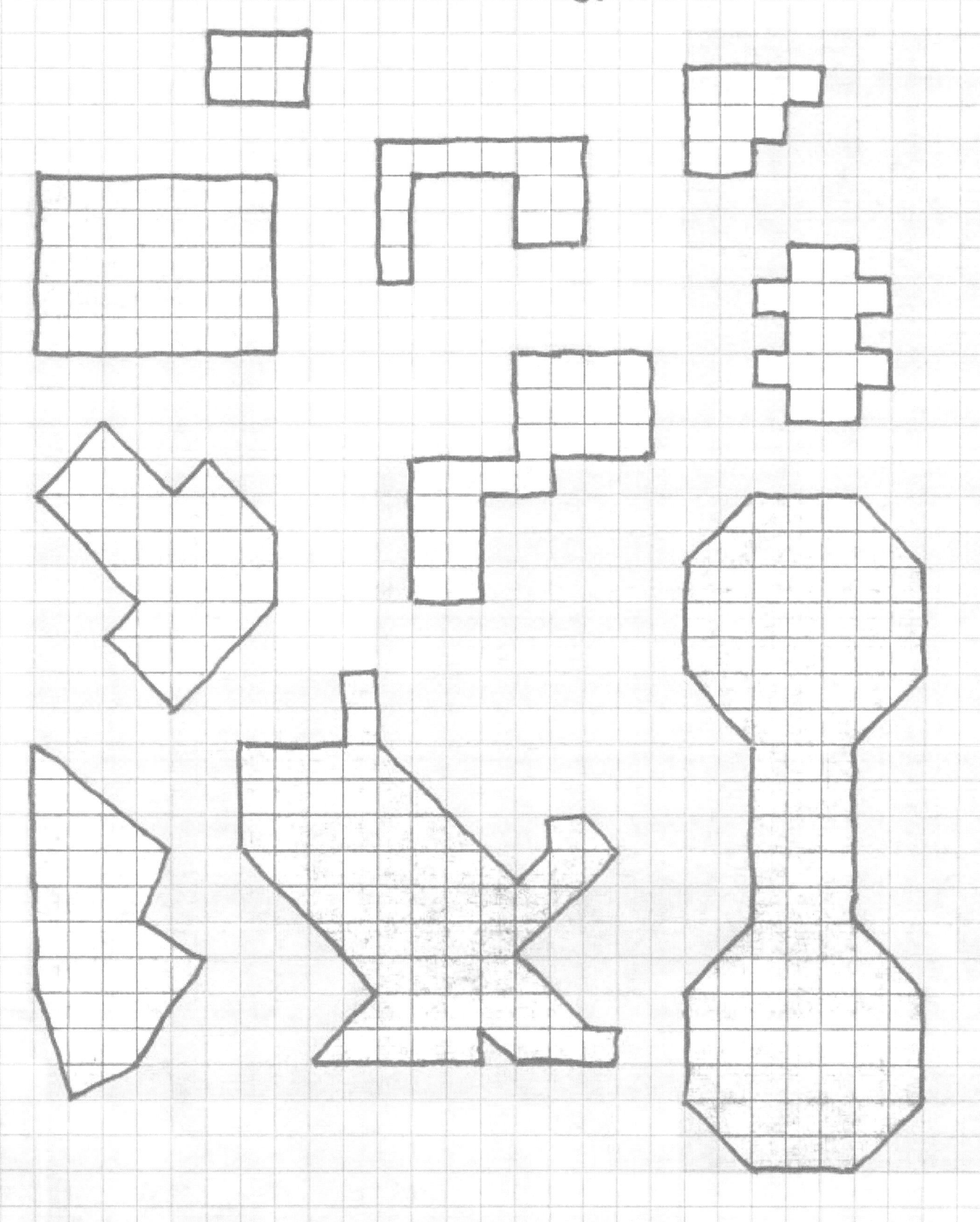

A DAY IN THE LIFE
OF A BIRD

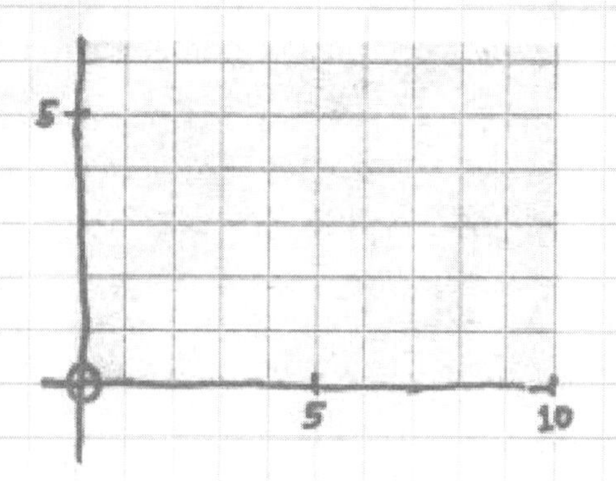

FROM O, MOVE 4 UNITS TO THE RIGHT AND 2 UNITS UP.
PUT A DOT THERE AND LABEL IT P. (AT 4,2)
FROM O, MOVE 5 UNITS TO THE RIGHT AND 2 UNITS UP.
PUT A DOT THERE AND LABEL IT Q. (AT 5,2)
FROM O, MOVE 5 UNITS TO THE RIGHT AND 3 UNITS UP.
PUT A DOT THERE AND LABEL IT R. (AT 5,3)
FROM O, MOVE 4 UNITS TO THE RIGHT AND 3 UNITS UP.
PUT A DOT THERE AND LABEL IT S. (AT 4,3)
DRAW A LINE FROM P TO Q TO R TO S TO P.
WE HAVE BUILT A SQUARE WITH AREA 1.

NEXT, PUT DOTS AT (7,1),(12,1),(12,4),(7,4)
LABEL THEM T,U,V,W, AND DRAW LINE TUVWT.
WE HAVE A RECTANGLE WITH AREA 15.

NEXT, DRAW △LMN WITH L AT (22,3), M AT (27,6)
AND N AT (23,8). WHAT IS ITS AREA?

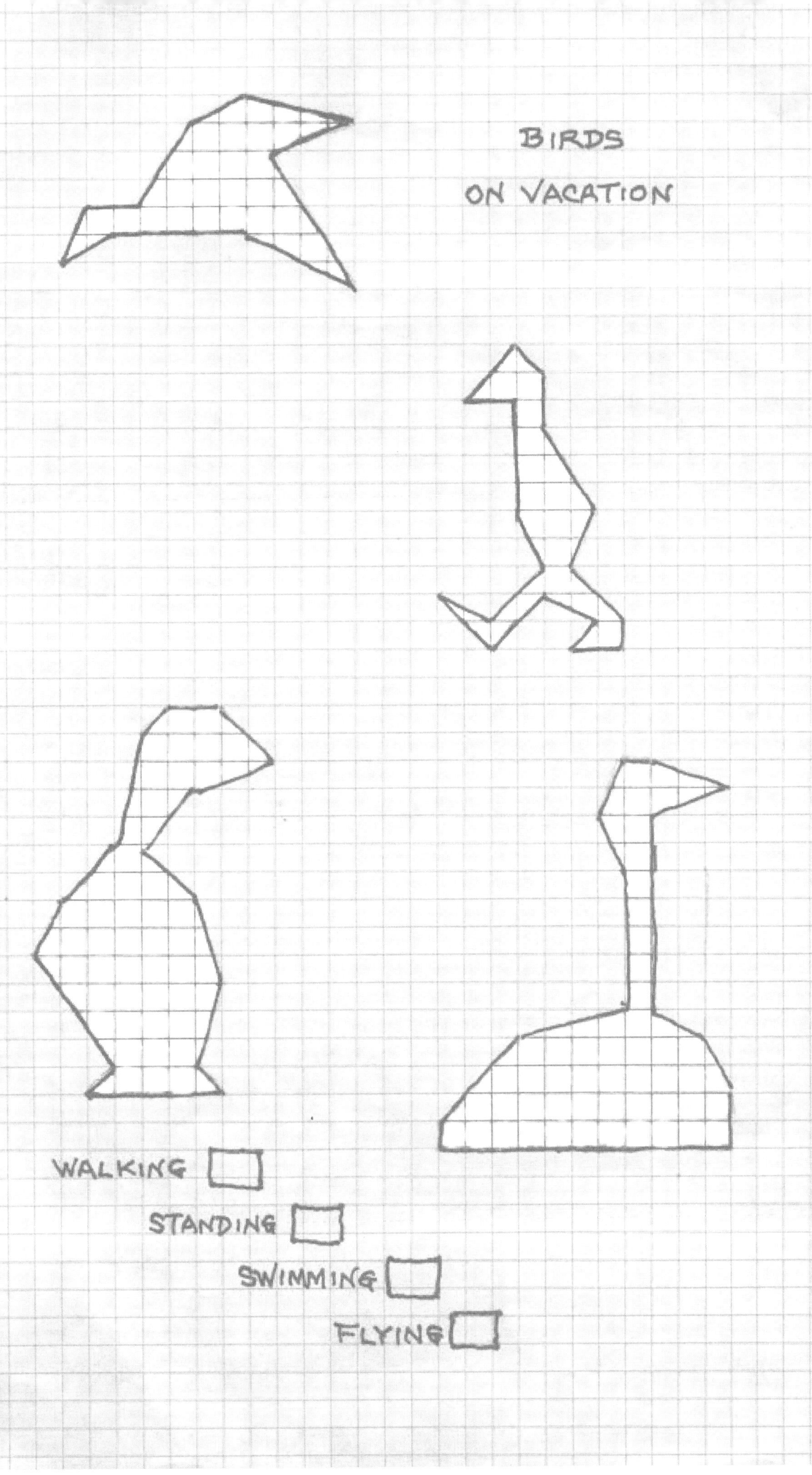
BIRDS
ON VACATION
WALKING
STANDING
SWIMMING
FLYING

IF EACH OF THESE

MARBLES

WEIGHS THE SAME

AND 5 MARBLES WEIGH 20 OUNCES

THEN 1 MARBLE WEIGHS [ ] OUNCES

IF ALL PAIRS OF TONGS WEIGH THE SAME,
AND 3 PAIRS OF TONGS WEIGH 11 POUNDS
THEN 7 PAIRS OF TONGS WEIGH [ ] POUNDS.

IF A GOAT CAN WALK 8 MILES IN 3 HOURS,
THEN IT CAN WALK [ ] MILES IN 1 HOUR,

AND [ ] MILES IN 5 HOURS.

IF A WORM CAN CRAWL 6 INCHES IN 5 MILLIDAYS,
THEN IT CAN CRAWL 1 INCH IN [ ] MILLIDAYS.

IF 1 COW CAN EAT 1 BALE OF HAY IN 7 DAYS,
THEN 2 COWS CAN EAT 1 BALE OF HAY IN [ ] DAYS.

4 SERGEANTS CAN PEEL 6 BINS OF POTATOES IN 10 HOURS;
2 SERGEANTS CAN PEEL 6 BINS IN [ ] HOURS,
2 SERGEANTS CAN PEEL 18 BINS IN [ ] HOURS
14 SERGEANTS CAN PEEL 18 BINS IN [ ] HOURS

IF 5 COWS CAN PEEL 7 BINS OF POTATOES IN D DAYS
1 COW CAN PEEL 7 BINS OF POTATOES IN [ ] DAYS
4 COWS CAN PEEL 7 BINS IN [ ] DAYS
4 COWS 1 BIN IN [ ] DAYS

IF 7 MOLES CAN DIG 3 HOLES IN 5 HOURS
. . . .
J MOLES CAN DIG K HOLES IN [ ] HOURS

PROBABILITY

WHEN AN ORDINARY COIN IS TOSSED,

THERE ARE TWO POSSIBLE OUTCOMES: TAIL AND HEAD.

AS A HEAD AND A TAIL ARE EQUALLY LIKELY,

THE PROBABILITY OF HEAD = $\frac{1}{2}$

AND THE PROBABILITY OF TAIL = $\frac{1}{2}$

WHEN THREE OF THESE COINS ARE SIMULTANEOUSLY TOSSED,

HOW MANY POSSIBLE OUTCOMES ARE THERE?

FIND THE PROBABILITIES OF EACH OUTCOME.

1) 3 HEADS

2)

⋮

WHEN AN EMPTY WATERING PAIL IS TOSSED,

THERE ARE THREE POSSIBLE OUTCOMES:

UPRIGHT, UP SIDE DOWN, ON THE CURVE.

ASSUME THAT THE PROBABILITIES OF THESE OUTCOMES

ARE $\frac{1}{7}$, $\frac{2}{7}$, AND $\frac{4}{7}$, RESPECTIVELY.

WHEN TWO IDENTICAL PAILS ARE TOSSED,

LIST EACH POSSIBLE OUTCOME, WITH ITS PROBABILITY.

WHEN TWO PAILS OF DIFFERENT SIZES ARE TOSSED,

LIST EACH POSSIBLE OUTCOME, WITH ITS PROBABILITY.

FOR A HARDER PROBLEM, CHANGE "TWO" TO "THREE."

(FOR THE SAKE OF YOUR NEIGHBORS, THESE LAST PROBLEMS SHOULD BE STUDIED IN A ROOM WITH A THICK RUG.)

PROBABILITY WITH COINS

WHEN A COIN IS FLIPPED,
THERE ARE TWO POSSIBLE OUTCOMES: HEADS AND TAILS.

WHEN A VERY THICK COIN IS FLIPPED, $p(H) = \frac{4}{7}$

$p(T) = \frac{2}{7}$

$p(\text{CURVE}) = \frac{1}{7}$

T
CURVE
H

$p(3H) =$

$p(3T) =$

$p(3C) =$

$p(2H, T) =$

$p(2H, C) =$

$p(H, 2T) =$

$p(H, 2C) =$

$p(T, 2C) =$

$p(H, T, C) =$

WHEN THREE OF THESE THICK COINS ARE TOSSED, FIND THE PROBABILITIES OF THE VARIOUS OUTCOMES.

WHY DON'T THESE PROBABILITIES ADD UP TO 1 ?

1) SAM REACHES INTO URN J WITHOUT LOOKING, AND PULLS OUT A MARBLE. SINCE THIS MARBLE IS JUST AS LIKELY TO BE RED AS TO BE BLUE, WE SAY: PROBABILITY (RED) $= \frac{1}{2}$
PROBABILITY (BLUE) $= \frac{1}{2}$

2) SALLY REACHES INTO URN K AND .... PROB. (RED) =
PROB. (GREEN) =

3) SENWOOD REACHES INTO URN L AND .. PROB. (RED) =
PROB. (GREEN) =
PROB. (BLUE) =

4) EMMA PICKS A MARBLE FROM URN M AND, WITHOUT LOOKING AT IT, DROPS IT INTO URN P. OLAF COMES ALONG, REACHES INTO URN P, AND PULLS OUT ONE OF THE 5 MARBLES. .... PROB. (RED) =
PROB. (GREEN) =
PROB. (BLUE) =

5) GIVEN URNS J, K, L, M, P AS ORIGINALLY DRAWN, TRANSFER A MARBLE FROM K TO P (NO PEEKING AT IT). THEN TRANSFER A MARBLE FROM M TO P. SHAKE URN CAREFULLY. NOW DRAW A MARBLE FROM URN P
(PROB) RED =
(PROB) GREEN =
(PROB) BLUE =

6) EMPTY URN L INTO THE WASTEBASKET. POUR THE CONTENTS OF URNS J, K, M, AND P INTO URN L. FIND THE PROBABILITY THAT TWO MARBLES DRAWN FROM URN L WILL BE OF <u>DIFFERENT</u> COLORS.

A PERSON WITH
THREE FINGERS ON EACH HAND

WOULD, FROM OUR POINT OF VIEW,
HAVE, ALL TOGETHER, SIX FINGERS.

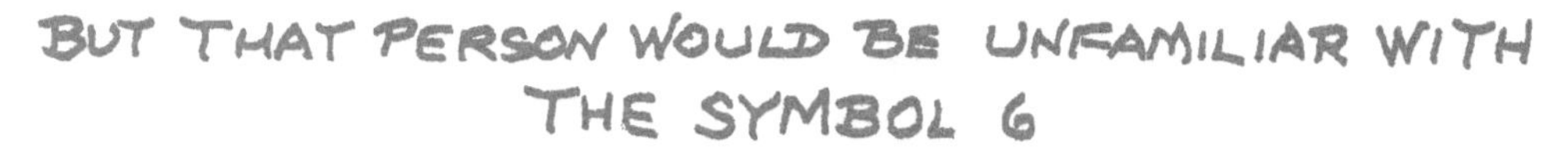

BUT THAT PERSON WOULD BE UNFAMILIAR WITH
THE SYMBOL 6

AND WOULD SAY: "NO, I HAVE 10 FINGERS, SEE, 1, 2, 3, 4, 5, 10!"

WE SAY THAT WE HAVE TEN FINGERS
BUT OUR FRIEND WOULD SAY:

"NO, YOU HAVE 14 FINGERS: SEE; 1, 2, 3, 4, 5, 10, 11, 12, 13, 14."

OUR FRIEND MIGHT HAVE OTHER EXPRESSIONS;

"TEN FOR ONE AND HALF A DOZEN FOR THE OTHER."

"TWO, FOUR, TEN, TWELVE: WHO(M) DO WE APPRECIATE?"
WHICH, HOWEVER, FAILS TO RHYME.

SOME ARITHMETIC PROBLEMS:

1) $4+2=\boxed{10}$

2) $15+1=\boxed{20}$

3) $3^2=\boxed{13}$

4) $(4)(5)=$

5) $2^{10}=2\cdot2\cdot2\cdot2\cdot2\cdot2$
$=12\cdot12=\boxed{144}$

10) $10^2=10\cdot10=\boxed{100}$

NOW, WITH EVEN FEWER FINGERS:

1) $2+2=\boxed{11}$

2) $(11)(11)=\boxed{121}$

10) $2^{10}=2\cdot2\cdot2=\boxed{22}$

11) $11+12=\boxed{}$

12) $10^2=\boxed{100}$

20) $\frac{1}{2}+\frac{1}{10}=\boxed{}$

SUPPOSE WE HAVE 2 FINGERS ON EACH HAND

WE WOULD COUNT 1, 2, 3, AND AFTER 3 WOULD BE 10.

THEN 11, 12, 13, 20, 21, 22, 23, 30, 31, 32, 33 .....

AND, SINCE 33 IS OUR HIGHEST TWO-DIGIT NUMBER, THE NEXT NUMBER WOULD BE ..... 100.

SO, $3 + 1 = 10$ $\quad 2 + 2 = 10$ $\quad 11 + 11 = 22$ $\quad 11 + 12 = 23$

$11 + 13 = 30$ $\quad 111 + 11 = 122$ $\quad 12 + 12 = 30$

HERE ARE 100 PROBLEMS:

1) $12 + 3 =$

2) $32 + 2 =$

3) $13 + 13 =$

10) $33 + 11 =$

11) $100 - 1 =$

12) $12 - 3 =$

13) $21 - 12 =$

20) $33 - 1 =$

21) $3 + 3 + 3 =$

22) $21 - 3 - 2 =$

23) $3^2 =$

30) $2W = 12$
$W =$

31) $3J = 12$
$J =$

32) $(33)^2 =$

33) $2^{11} =$

100) $12! =$

1) $1 + 1 = \boxed{10}$

10) $\frac{1}{10} + \frac{1}{10} = \frac{10}{10} = \boxed{1}$

11) $10^{10} =$

100) $11 + 11 =$

101) $100 + 1 =$

110) $\frac{1}{10} + \frac{1}{11} =$

111) $(11)^{10} =$

1000) $(11)^{11} =$

HOW DOES SOMEONE COUNT

WHO HAS MORE FINGERS THAN WE ?

1, 2, 3, 4, 5, 6, 7, 8, 9, , , 10, 11, 12, 13, ...

1) $(3)(4) =$

2) $6 + 6 =$

3) $9 + 3 =$

4) $9 + 1 =$

5) $4^2 = (4)(4) =$

6) $(5)(5) =$

THIS PERSON NEEDS TWO STRANGE NEW SINGLE-DIGIT SYMBOLS

CHOOSE YOUR SYMBOLS

THE ONE ON THE LEFT WILL BE THE ANSWER TO PROBLEM 4 AND IT WILL GO HERE →

THE ONE ON THE RIGHT WILL GO AFTER THE 2 IN PROBLEM 7 AND IT WILL GO HERE →

7) $(5)(7) = \boxed{2\ \ }$

8) $2^4 = \boxed{14}$

9) $2^6 =$

) $8 \cdot 9 = \boxed{60}$

) $100 - 1 =$

10) $57 + 66 =$

	1	2	3	4	5	6	10
1	1	2	3	4	5	6	10
2	2			11			
3	3						
4	4						
5	5	13		26			
6	6						
10	10						100

MULTIPLICATION TABLES

USED BY

OUR

FRIENDS

	1	2	3	4	5	6	7	10
1	1	2	3	4	5	6	7	10
2	2							
3	3							
4	4							
5	5							
6	6							
7	7							
10	10							

Z E R O

IS, IN SOME WAYS, A NUMBER,
AND, IN OTHER WAYS, <u>NOT</u> A NUMBER

WHEN WE ADD ZERO TO A NUMBER, NOTHING HAPPENS.
THE NUMBER DOES NOT CHANGE. $5 + 0 = 5$

WHEN WE SUBTRACT ZERO FROM A NUMBER, NOTHING HAPPENS.
THE NUMBER DOES NOT CHANGE. $7 - 0 = 7$

WHEN WE MULTIPLY A NUMBER BY ZERO, THE RESULT IS ZERO.
$6(0) = 0 + 0 + 0 + 0 + 0 + 0 = 0$

WHEN WE DIVIDE ZERO BY A NON-ZERO NUMBER, THE RESULT IS ZERO.
$0 \div 3 = 0$, BECAUSE 3 TIMES $0 = 0$

<u>BUT</u>, WHAT HAPPENS WHEN WE TRY TO DIVIDE A NUMBER BY ZERO??

LET US REMEMBER WHAT DIVISION MEANS:
14 DIVIDED BY 2 MEANS: WHAT DO WE MULTIPLY 2 BY TO GET 14?

BUT WHAT IS 5 DIVIDED BY ZERO?
WHAT DO WE MULTIPLY ZERO BY TO GET 5?
IN OTHER WORDS: HOW MANY ZEROES MUST WE ADD TO GET 5?

LET'S TRY 3

$$\begin{array}{r} 0 \\ 0 \\ +\,0 \\ \hline 0 \end{array}$$

NO

LET'S TRY 5

$$\begin{array}{r} 0 \\ 0 \\ 0 \\ 0 \\ +\,0 \\ \hline 0 \end{array}$$

NOPE!

00000
0
0
0000
0
0
0
0000
0

INTERESTING IDEA,

BUT IT <u>STILL</u> ADDS UP ONLY TO ZERO.

TRY ANOTHER WAY:
$0 + 0 + 0 + 0 + 0 + 0 + 0 + 0 + 0 + 0 + 0 + 0 + 0 + 0 + 0 + 0 + 0 + 0 + 0 + 0 +$
IT BECOMES CLEAR THAT THERE IS NO ANSWER TO THIS PROBLEM.
$5 \div 0$ HAS NO ANSWER.
LIKEWISE, $7 \div 0$ HAS NO ANSWER.
IN FACT, THERE IS NO NON-ZERO NUMBER WHICH CAN BE DIVIDED BY ZERO

WHAT IS ZERO DIVIDED BY ZERO?

OR

HOW MANY ZEROES MUST WE ADD TO GET ZERO?

AS WE SAW ON THE PREVIOUS PAGE: 3 ZEROES ADD UP TO ZERO.
5 ZEROES ADD UP TO ZERO.
18 ZEROES ADD UP TO ZERO.
284 ZEROES ADD UP TO ZERO.

SO, IF ZERO DIVIDED BY ZERO EQUALS ANYTHING,
IT EQUALS 3, 5, 18, 284, AND EVERY OTHER NUMBER.
BUT NO NUMBER CAN EQUAL A LOT OF DIFFERENT NUMBERS.

WORSE YET, CONSIDER THE FOLLOWING PATTERNS:

$\frac{1}{1}=1$, $\frac{2}{2}=1$, $\frac{3}{3}=1$, $\frac{a}{a}=1$; BY THIS PATTERN $\frac{0}{0}=1$.

BUT $\frac{0}{1}=0$, $\frac{0}{2}=0$, $\frac{0}{3}=0$, $\frac{0}{b}=0$; BY THIS PATTERN $\frac{0}{0}=0$.

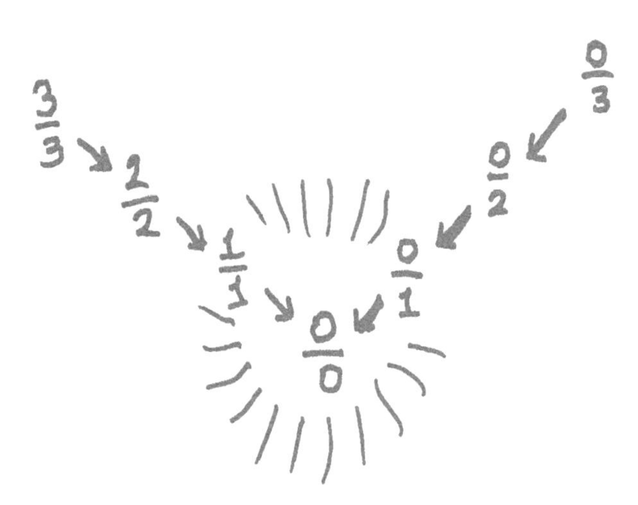

TROUBLE! NO NUMBER CAN BE DIVIDED BY ZERO,

AND WE REFUSE
TO DISCUSS THE MATTER!

e

IF, ON 1 JANUARY,

WE INVEST $1. IN A BANK WHICH PAYS US 100% PER YEAR ON OUR INVESTMENT, THEN AFTER A YEAR, ON THE FOLLOWING 1 JANUARY, WE COLLECT FROM THE BANK OUR $1. PLUS THE $1. INTEREST, MAKING A TOTAL OF $2.

IF, ON 1 JANUARY,

WE INVEST $1. IN ANOTHER BANK WHICH PAYS US 100% PER YEAR BUT PAYS THE INTEREST <u>TWICE</u> A YEAR, THE FIRST TIME ON 1 JULY AND THE SECOND TIME ON THE FOLLOWING 1 JANUARY, LET'S SEE WHAT HAPPENS:

ON 1 JULY OUR $1. HAS EARNED 50¢. IF WE LEAVE THIS 50¢ IN THE BANK WITH THE ORIGINAL DOLLAR, THEN OUR INVESTMENT OF $1.50 WILL, ON THE FOLLOWING 1 JANUARY, HAVE EARNED FOR US 50% OF ITSELF, WHICH IS 75¢, MAKING A TOTAL OF $2.25. IN OTHER WORDS THE BANK WILL HAVE MULTIPLIED OUR $1. TWICE BY $1+\frac{1}{2}$, GIVING US FIRST (ON 1 JULY) $1.50 AND THEN (ON THE FOLLOWING 1 JANUARY) $2.25.

IF, ON 1 JANUARY,

WE FIND A BANK WHICH PAYS 100% PER YEAR AND COMPOUNDS THE INTEREST NOT JUST ONCE, NOR TWICE, BUT FOUR TIMES DURING THE YEAR, THEN OUR DOLLAR WILL BE MULTIPLIED ON 1 APRIL BY $1+\frac{1}{4}$, ON 1 JULY BY $1+\frac{1}{4}$, ON 1 OCTOBER BY $1+\frac{1}{4}$, <u>AND</u>, ON THE FOLLOWING 1 JANUARY BY $1+\frac{1}{4}$, GIVING US ON 1 JANUARY A TOTAL OF:

1 TIMES $1\frac{1}{4}$ TIMES $1\frac{1}{4}$ TIMES $1\frac{1}{4}$ TIMES $1\frac{1}{4}$

OR $\left(1\frac{1}{4}\right)^4 = \left(\frac{5}{4}\right)^4 = \frac{625}{256} = \$2.44^+$

CONTINUING OUR SEARCH

FOR A BANK WHICH PAYS 100% PER YEAR BUT COMPOUNDS MORE OFTEN, WE FIND A BANK WHICH PAYS 100% PER YEAR AND COMPOUNDS 12 TIMES. INVESTING OUR DOLLAR HERE WILL, AFTER A YEAR, GIVE US:

$$1 \cdot 1\tfrac{1}{12} \cdot 1\tfrac{1}{12} \cdot 1\tfrac{1}{12} \cdot 1\tfrac{1}{12} \cdot 1\tfrac{1}{12} \cdot 1\tfrac{1}{12} \cdot 1\tfrac{1}{12} \cdot 1\tfrac{1}{12} \cdot 1\tfrac{1}{12} \cdot 1\tfrac{1}{12} \cdot 1\tfrac{1}{12} \cdot 1\tfrac{1}{12} = \left(1+\tfrac{1}{12}\right)^{12}.$$

THIS WORKS OUT TO $2.613.⁺

TABULATING THESE RESULTS

NUMBER OF TIMES DURING THE YEAR THAT THE INTEREST IS COMPOUNDED	VALUE OF INVESTMENT ON THE FOLLOWING 1 JANUARY
1	$\left(1+\frac{1}{1}\right)^{1} = 2$
2	$\left(1+\frac{1}{2}\right)^{2} = 2.25$
4	$\left(1+\frac{1}{4}\right)^{4} = 2.44^{+}$
12	$\left(1+\frac{1}{12}\right)^{12} = 2.613^{+}$
100	$\left(1+\frac{1}{100}\right)^{100} = 2.7048^{+}$
1000	$\left(1+\frac{1}{1000}\right)^{1000} = 2.7169^{+}$
1000000	$\left(1+\frac{1}{1000000}\right)^{1000000} = 2.71828046^{+}$
1000000000	$\left(1+\frac{1}{1000000000}\right)^{1000000000} = 2.718281827^{+}$

WE SEE THAT THIS LAST VALUE DIFFERS
FROM THE PREVIOUS ONE
IN THE SIXTH DIGIT AFTER THE DECIMAL.

NO MATTER HOW OFTEN THE INTEREST IS COMPOUNDED,

THE VALUE OF THE INVESTMENT AFTER ONE YEAR
WILL NEVER REACH 3.
NOR WILL IT REACH 2.8, OR EVEN 2.72.

THERE ARE, OF COURSE,
MANY NUMBERS WHICH THE VALUE WILL NEVER REACH,

AND THE SMALLEST OF THESE IS CALLED **e**.

e IS LARGER THAN 2.718281828
BUT SMALLER THAN 2.718281829

RECURSION

IS AN AMUSING APPROACH TO PROBLEMS WHICH APPEAR TO GO ON FOREVER.

TAKE THE REPEATING DECIMAL .33333 →
TO FIND WHAT FRACTION THIS EQUALS,
CALL THAT FRACTION W.

THEN $W = .33333 \rightarrow$

MULTIPLY BOTH SIDES OF THE EQUATION BY 10.

$10W = 3.33333 \rightarrow$

OR $10W = 3 + .33333 \rightarrow$

OR $10W = 3 + W$

SUBTRACTING W FROM EACH SIDE GIVES US

$9W = 3$

SO $\boxed{W = \frac{3}{9} = \frac{1}{3}}$

TAKE .272727 →

$W = .272727 \rightarrow$

$100W = 27.272727 \rightarrow$

$100W = 27 + W$

$99W = 27$

$11W = 3$

$\boxed{W = \frac{3}{11}}$

TAKE .4373737 ....

$W = .4373737 \ldots$

$100W = 43.7373737 \ldots$

$W = .4373737$

$99W = 43.3$

$990W = 433$

$\boxed{W = \frac{433}{990}}$

TAKE $1 + \frac{2}{5} + \frac{4}{25} + \frac{8}{125} + \ldots$

MULTIPLY BOTH SIDES BY $\frac{2}{5}$

$W = 1 + \frac{2}{5} + \frac{4}{25} + \frac{8}{125} + \ldots$

$\frac{2}{5}W = \frac{2}{5} + \frac{4}{25} + \frac{8}{125} + \ldots$

ADD 1 TO EACH SIDE:

$\frac{2}{5}W + 1 = 1 + \frac{2}{5} + \frac{4}{25} + \frac{8}{125} + \ldots$

OR

$\frac{2}{5}W + 1 = W$

$2W + 5 = 5W$

$3W = 5$

$\boxed{W = 1\frac{2}{3}}$

FIND THE FRACTIONS WHICH REPRESENT THE FOLLOWING DECIMALS

1) $0.7777 \rightarrow$

2) $0.3939393 \rightarrow$

3) $0.166666 \rightarrow$

4) $0.363636 \rightarrow$

5) $0.363363363363 \rightarrow$

FIND THE VALUE OF:

6) $1 + \cfrac{1}{2 + \cfrac{1}{1 + \cfrac{1}{2 + \cdots}}}$

7) $1 + \frac{2}{3} + \frac{4}{9} + \frac{8}{27} + \frac{16}{81} + \cdots$

8) $3 + \sqrt{3 + \sqrt{3 + \sqrt{3 + \cdots}}}$

RESIDUE CLASSES

ON THE TOP ROW WE WRITE EVERY MULTIPLE OF 7

..... -35 -28 -21 -14 -7 0 7 14 21 28 35 42 ...

ON THE SECOND ROW WE PUT THE NUMBERS WE GET BY ADDING +1 TO THE NUMBERS IN THE TOP ROW

..... -34 -27 -20 -13 -6 1 8 15 22 29 36 43 ....

ADDING +1 AGAIN GIVES US

..... -42 -35 -28 -21 -14 -7 0 7 14 21 28 35 42 ....
..... -41 -34 -27 -20 -13 -6 1 8 15 22 29 36 43 ....
..... -40 -33 -26 -19 -12 -5 2 9 16 23 30 37 44 ....
..... -39 -32 -25 -18 -11 -4 3 10 17 24 31 38 45 ....
..... -38 -31 -24 -17 -10 -3 4 11 18 25 32 39 46 ....
..... -37 -30 -23 -16 -9 -2 5 12 19 26 33 40 47 ....
..... -36 -29 -22 -15 -8 -1 6 13 20 27 34 41 48 ....

OF COURSE WE DO NOT NEED TO PUT ANOTHER ROW, SINCE IT WOULD CONTAIN THE NUMBERS WHICH ARE ALREADY IN THE TOP ROW

EVERY WHOLE NUMBER, POSITIVE OR NEGATIVE, IS IN ONE OF THE ROWS. FOR EXAMPLE, 815 IS IN ROW ③, 57 IS IN ROW ①, ETC....

NOW LET US SEE WHAT HAPPENS WHEN WE ADD A NUMBER IN ROW ① TO A NUMBER IN ROW ②. WELL, 15 IN ROW ① ADDED TO 23 IN ROW ② GIVES US 38 IN ROW ③. NOW TAKE A NUMBER FROM ROW ④, SAY 18, AND ADD IT TO ANOTHER NUMBER, IN ROW ⑤, SAY 26, AND WE GET 44, WHICH IS IN ROW ②.

SO WE WRITE ① + ② = ③ AND ④ + ⑤ = ②.

LIKEWISE

③ + ④ = ⓪ , 2 + 6 = 1 , 5 + 6 = 4, ETC.

RESIDUE CLASS ARITHMETIC IN J_7

ADDING) $1+2=3$ $\quad 2+3=5$ $\quad 4+5=2$ $\quad 6+0=6$

$2+4=6$ $\quad 5+6=4$ $\quad 1+6=0$ $\quad 4+4=1$

SUBTRACTING) $5-2=3$ $\quad 6-1=5$ $\quad 2-3=6$ $\quad 4-4=0$

$2-5=4$ $\quad 1-6=2$ $\quad 5-0=5$ $\quad 0-5=2$

MULTIPLYING) $4(2)=1$ $\quad 2(5)=3$ $\quad 2\cdot 2=4$

$3(5)=1$ $\quad 3(3)=2$ $\quad 4\cdot 4=2$

$4(6)=3$

THAT'S A SWITCH!

DIVIDING) $\frac{1}{2}=4$ $\quad \frac{1}{3}=5$ $\quad \frac{1}{4}=2$ $\quad \frac{1}{5}=3$

$\frac{2}{3}=3$ $\quad \frac{2}{5}=6$ $\quad \frac{3}{4}=6$

$W^2=2$ GIVES US $\begin{cases} W=3 \\ W=4 \end{cases}$

$W^2=1$ GIVES US $\begin{cases} W=1 \\ W=6 \end{cases}$

$W^2=4$ GIVES US $\begin{cases} W=2 \\ W=5 \end{cases}$

$W^2=5$ NO ANSWER

$W^2=6$ NO ANSWER

$W^2=3$ NO ANSWER

IN J_5

RESIDUE CLASS ARITHMETIC

ADDING) $1+2=3$ $2+3=0$ $1+3=4$

$2+4=1$ $3+4=2$ $4+4=3$

SOLVE THE FOLLOWING IN MODULO 5

$(4)(4) =$

$(2)(3)(4) =$

$3^3 =$

$W^3 = 2$

$W =$

$[1 + (2)(2) + 4]3 =$

$(4-1)(3-1)$

$W^2 = 4$

$2^4 =$

$\sqrt{W} = 1$

$W^2 = 2$

$W^2 = 3$

$W^2 = 1$

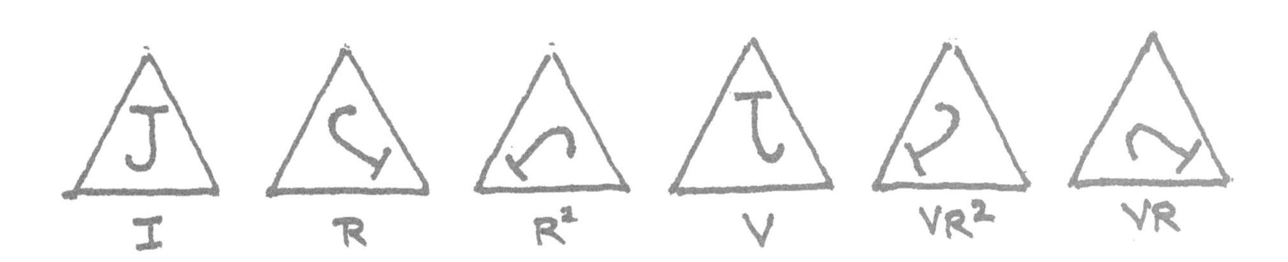

IF WE PICK UP DIAGRAM I AND ROTATE IT $\frac{1}{3}$ TURN CLOCKWISE,
WE CAN CALL THAT OPERATION R

IF WE PICK UP DIAGRAM I AND PERFORM OPERATION R
AND THEN PERFORM OPERATION R AGAIN,
WE CAN CALL THAT OPERATION R^2.

DO YOU SEE THAT IF WE PERFORM R THREE TIMES, THE J WILL
BE BACK IN ITS ORIGINAL UPRIGHT POSITION? THAT IS: RRR = I

IF WE PERFORM R THEN V, WHICH OF THE SIX POSITIONS
HAVE WE REACHED? IT'S NOT VR!!

DO YOU SEE THAT $RRR = R^3$ IS EQUAL TO I, SINCE THE J WILL
BE BACK IN ITS ORIGINAL POSITION?

DO YOU SEE WHY $V^2 = I$?

DO YOU SEE WHY RV = NOT VR, BUT VR^2?

CAN YOU FIND A ROTATION Q OF A DIE
SUCH THAT $Q^4 = I$?

CAN YOU FIND A ROTATION W OF A DIE
SUCH THAT $W^3 = I$?

STOP → (I) STOP

STOP → (R) STOP

STOP → (R^2) STOP

STOP → (R^3) STOP

STOP → (V) STOP

STOP → (VR) STOP

STOP → (VR^2) STOP

STOP → (VR^3) STOP

IF WE ROTATE THE STOP SIGN I 1/4 TURN CLOCKWISE, WE GET R.

ROTATING AGAIN, WE GET R^2

ROTATING ONCE MORE, WE GET RRR, OR R^3.

AND ROTATING IT YET AGAIN, WE FIND OURSELVES BACK WHERE WE STARTED.

SO $R^4 = I$.

IF WE SPIN THE SIGN AROUND A VERTICAL AXIS, WE GET V.

IF WE SPIN AND THEN ROTATE, WE GET VR.

IF WE SPIN TWICE, WE GET $V^2 =$

IF WE SPIN, THEN ROTATE TWICE, WE GET VRR, OR VR^2

IF WE ROTATE, THEN SPIN, WHICH OF THE EIGHT DO WE GET?

RV =

R^7 = RRRRRRR =

VRV =

$(VR)^3$ = VRVRVR =

SQUOMINOES

1) THERE IS, OF COURSE, ONLY ONE UNIMO:

2) AND THERE IS ONLY ONE DOMINO:

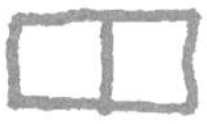

3) BUT THERE ARE TWO TROMINOES:

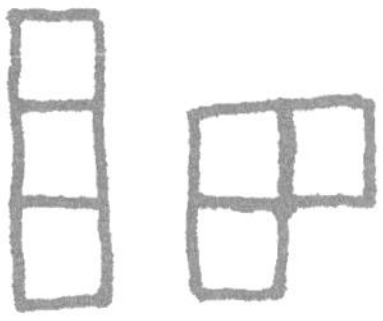

4) AND 5 QUADROMINOES:

CAN YOU FIND THE OTHER THREE?

(LET US NOT COUNT MIRROR IMAGES.)

5) AND 12 QUINTOMINOES:

6) HOW MANY HEXOMINOES?

TRINIMOES

1)

2)

3)

4)

5)

6)

7)

8)

9)

10)

HEXIJOES

1)

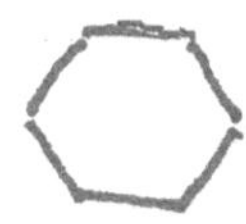

2)

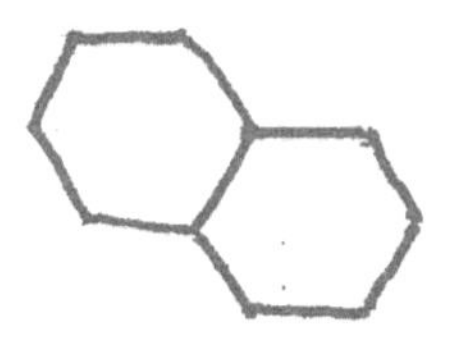

3)

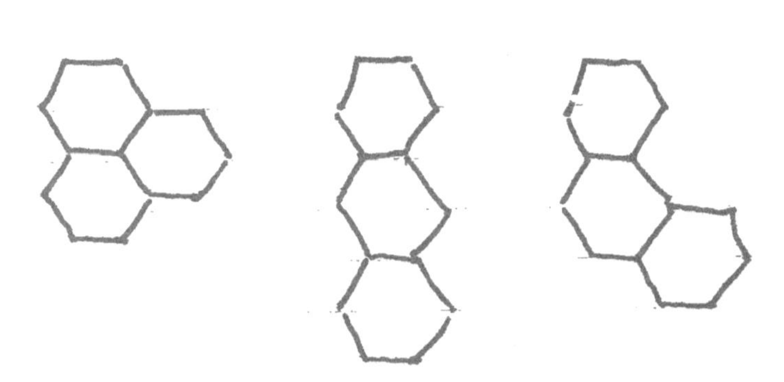

4)

IS THERE AN OBLONG PROBLEM ?

IF THERE IS
SOMETIMES
AN OBLONG
PROBLEM,

EXPLAIN WHEN, HOW,
AND WHY.

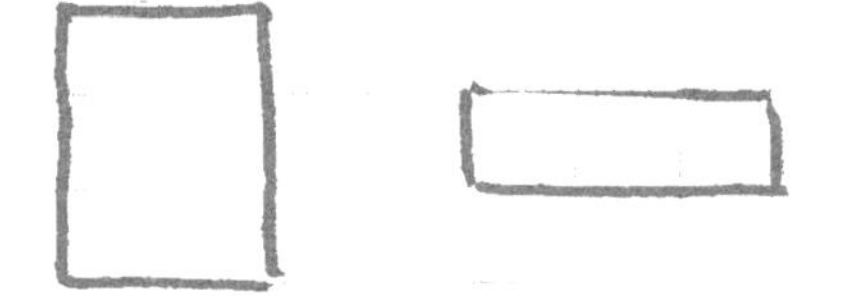

IS THERE A
CIRCLE
PROBLEM ?

IS THERE A
PENTOMINO
PROBLEM ?

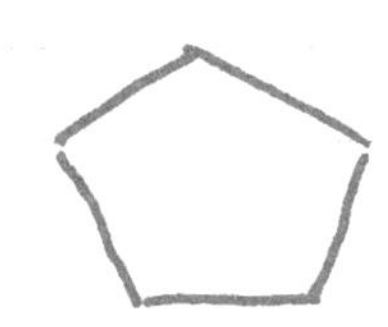

10 IS A NUMBER WHICH EVERYONE UNDERSTANDS.

10^{10}, WHICH IS 10,000,000,000, OR TEN BILLION, IS A NUMBER WHICH MAY APPEAR IN SCIENCE OR FINANCE, AND WE MAY RECOGNIZE, IF NOT UNDERSTAND, THIS NUMBER.

$10^{10^{10}}$, WHICH, BY AGREEMENT, MEANS NOT $(10^{10})^{10}$, WHICH IS ONLY 10^{100}, BUT $10^{(10^{10})}$.

THIS NUMBER IS 1 WITH TEN BILLION ZEROES AFTER IT.

PUTTING FIVE ZEROES PER INCH,
THIS NUMBER, WRITTEN OUT,
WOULD EXTEND ABOUT ☐ MILES.

AND WHAT ABOUT 10 TO THIS POWER, WRITTEN $10^{10^{10^{10}}}$?

THIS MEANS $10^{\left[10^{(10^{10})}\right]}$

OR $10^{(10^{10000000000})}$.

THIS NUMBER IS BEYOND THE IMAGINATION
OF THE AVERAGE PERSON.
WRITTEN OUT, IT WOULD EXTEND
ABOUT ☐
LIGHT YEARS.

SOME LARGER NUMBERS

WHICH OF THE NUMBERS BELOW IS BY FAR THE LARGEST?

$$100^{100} \qquad 50^{50} \qquad 20^{(20^{20})} \qquad \left[\left(10^{10}\right)^{10}\right]^{10} \qquad 2^{\left\{2^{\left[2^{\left(2^{2}\right)}\right]}\right\}}$$

TO ATTACK THIS PROBLEM,
WE TRY TO CALCULATE THE NUMBER OF DIGITS IN EACH ANSWER.

50^{50} IS CLEARLY LESS THAN 100^{100}.
SO WE FORGET 50^{50}.

$$\left[\left(10^{10}\right)^{10}\right]^{10} = \left[10^{100}\right]^{10} = 10^{1000}$$

$$2^{\left\{2^{\left[2^{\left(2^{2}\right)}\right]}\right\}} = 2^{\left\{2^{\left[2^{4}\right]}\right\}} = 2^{\left\{2^{16}\right\}} = 2^{65536}$$

THIS IS CERTAINLY LESS THAN 10^{65536}.

WHAT ABOUT $20^{(20^{20})}$?

$$20^{(20^{20})} = 20^{(10^{20} \cdot 2^{20})} = 20^{104857600000000000000000000}$$

TRANSFINITE

YOU ARE THE MANAGER OF A HOTEL. IT IS A VERY LARGE HOTEL. ON THE DOOR OF THE FIRST ROOM IS A 1. THE NEXT ROOM HAS A 2 ON THE DOOR. THE HOTEL IS SO LARGE THAT FOR EVERY POSITIVE WHOLE NUMBER THERE IS A ROOM WITH THAT NUMBER ON THE DOOR. SO THERE IS A ROOM 967, ROOM 8388608, ROOM $10^{53} + 5968$, JUST TO MENTION A FEW.

ON ONE DARK STORMY NIGHT YOUR HOTEL IS FULL, MEANING THAT IN EACH ROOM THERE IS EXACTLY ONE GUEST. SUDDENLY HORSES' HOOFS ARE HEARD, AND INTO YOUR HOTEL COMES A TIRED WET STRANGER, WHO ASKS IF YOU HAVE AN EMPTY ROOM AVAILABLE. AFTER YOU SAY NO, THE STRANGER WHISPERS A FEW WORDS INTO YOUR EAR, AFTER WHICH YOU MOVE TO YOUR LOUDSPEAKER AND MAKE THE FOLLOWING ANNOUNCEMENT:

"NOW HEAR THIS: WILL EACH GUEST MOVE IMMEDIATELY FROM YOUR ROOM TO THE ROOM WHICH HAS THE NEXT NUMBER HIGHER THAN THE NUMBER OF YOUR PRESENT ROOM."

SO AFTER THIS ANNOUNCEMENT THE GUEST IN ROOM 1 MOVES TO ROOM 2, THE GUEST IN ROOM 2 MOVES TO ROOM 3. THE GUEST IN ROOM 967 MOVES TO ROOM 968, AND SO ON. YOU THEN TELL THE STRANGER THAT ROOM 1 WILL, IN A MOMENT, BE AVAILABLE.

SINCE THERE IS NO HIGHEST NUMBER, THERE IS NO LAST ROOM, SO EVERY GUEST WILL NOW BE ACCOMMODATED.

LET US NOW LOOK AT THE NUMBERS: WE SHALL REPRESENT THE NUMBER OF (POSITIVE WHOLE) NUMBERS BY D. THEN THERE ARE D ROOMS IN THE HOTEL. AND BEFORE THE ARRIVAL OF THE STRANGER THERE WERE D GUESTS. WITH THE STRANGER, THERE ARE NOW $D+1$ GUESTS.

IT IS CLEAR THAT AFTER THE STRANGER'S ARRIVAL EACH ROOM STILL HAS EXACTLY ONE GUEST. AND SINCE THERE ARE NOW $D+1$ GUESTS IN THE HOTEL, OCCUPYING D ROOMS, THEN THE NUMBER OF GUESTS EQUALS THE NUMBER OF ROOMS. HENCE WE SEE THE REMARKABLE EQUATION

$$D + 1 = D$$

JUST AS A NUMBER, FOR EXAMPLE 7, IS NOT AN OBJECT BUT AN ABSTRACT IDEA CONCERNING HOW MANY OBJECTS WE MAY BE THINKING OF, D, BEING THE NUMBER OF NUMBERS, IS AN ABSTRACTION OF AN ABSTRACTION.

HOW DOES D BEHAVE?

IF WE SUBTRACT D FROM EACH SIDE OF THE EQUATION ABOVE, WE GET $1 = 0$, WHICH IS FALSE. HENCE D CANNOT BE SUBTRACTED.

SO D CANNOT BE SUBTRACTED

WE NOW SUPPOSE THAT THE HOTEL IS FULL,
THAT IS, 1 GUEST IN EACH OF D ROOMS
MAKING D GUESTS IN ALL.

THIS TIME EACH GUEST DECIDES TO ASK FOR TWO ROOMS,
INSTEAD OF JUST ONE.

SURELY THIS IS NOT POSSIBLE! BUT IT IS!

"NOW HEAR THIS! EACH GUEST WILL MOVE AT ONCE
TO THE ROOM WHICH HAS THE NUMBER
WHICH IS TWICE THE NUMBER OF YOUR PRESENT ROOM.
YOU MAY HAVE THAT ROOM AND THE ROOM
WITH THE NEXT SMALLER NUMBER."

SO THE GUEST IN ROOM 7 MOVES INTO ROOMS 14 AND 13.
THE GUEST IN ROOM 568 MOVES INTO ROOMS 1136 AND 1135.
AND EVERY GUEST WILL GET TWO ROOMS.

IF EACH OF D GUESTS GETS 2 ROOMS,
THAT MEANS THAT THE HOTEL HAS 2D ROOMS.
BUT THE HOTEL DIDN'T ADD ANY ROOMS;
THERE ARE STILL D ROOMS.
SO WE SEE ANOTHER EQUATION: $2D = D$
IF WE DIVIDE BOTH SIDES OF THIS EQUATION BY D,
(WHICH IS CERTAINLY NOT ZERO),
WE GET $2 = 1$, WHICH IS FALSE.
HENCE WE CANNOT DIVIDE BY D.

D IS THE SMALLEST TRANSFINITE NUMBER.

HOW MANY LARGER ONES ARE THERE??

TONGS

1) IF 3 PUPPIES OF A LITTER WEIGH 6 OUNCES, EACH PUPPY PROBABLY WEIGHS ABOUT ☐ OUNCES.

2) IF YOU CAN WALK A MILE IN ABOUT 20 MINUTES, YOU CAN PROBABLY WALK 5 MILES IN ABOUT ☐

3) IF YOU CAN PEEL 17 APPLES IN ABOUT 5 MINUTES, YOU CAN PEEL 30 APPLES IN ABOUT ☐ MINUTES.

4) IF AXEL CAN EARN J DOLLARS IN K DAYS, HOW LONG WILL TAKE AXEL TO EARN 3J DOLLARS? ☐

5) IF 8 POMEGRANATES WEIGH J OUNCES, HOW MUCH DO 15 POMEGRANATE PROBABLY WEIGH? ☐ OUNCES

6) IF 3 TONGS WEIGH 11 POUNDS, HOW MUCH DO 11 TONGS WEIGH? ☐ OUNCES

7) IF FRANK CAN WALK M FEET IN Q MINUTES, HOW FAR CAN FRANK WALK IN 3 MINUTES? ☐

8) IF A PORCUPINE CAN CRAWL 7 YARDS IN 20 SECONDS, HOW MANY FEET CAN IT CRAWL IN 9 MINUTES? ☐

9) A CIRCULAR FIELD IS ENCLOSED BY A FENCE. THE HORSE IN THIS FIELD HAS LOTS OF GRASS TO EAT. THE HORSE LIKES THE AMOUNT OF GRASS HE CAN REACH, BUT WISHES THE FIELD WERE SQUARE INSTEAD OF CIRCULAR. SQUARE 1 IS TOO SMALL, SQUARE 6 TOO LARGE.

BEST GUESS ☐

1 2 3 4 5

SUPPOSE WE HAVE A POLE
FIXED VERTICALLY IN THE WATER.

SUPPOSE FURTHER THAT WE SEE
A WATERBUG SITTING ON THE POLE
AT POSITION 3.

AND SUPPOSE THAT THIS SORT OF WATERBUG
CANNOT SWIM OR SEPARATE ITSELF FROM THE POLE.

SO THE ONLY WAY IT CAN MOVE
IS BY CRAWLING UP OR DOWN.

ONE MORNING WE SEE IT AT POSITION 3.

THAT AFTERNOON WE SEE IT AT POSITION −4.

WE KNOW, THEREFORE, WITHOUT BEING TOLD,
THAT THERE MUST HAVE BEEN
AT LEAST ONE INTERMEDIATE INSTANT
AT WHICH IT WAS AT POSITION −2.

THIS IS CALLED THE PRINCIPLE OF CONTINUITY.

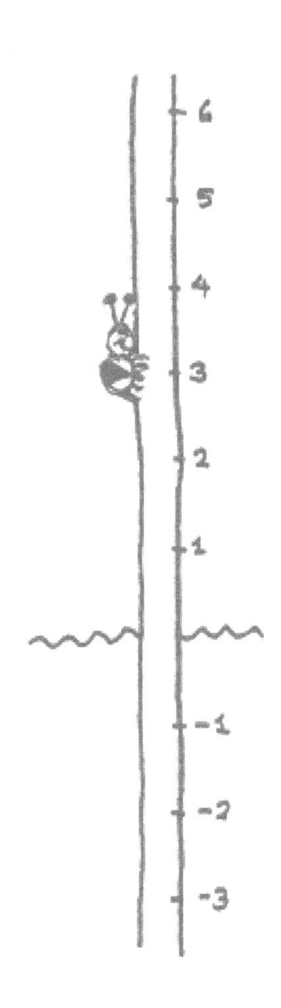

CONTINUITY

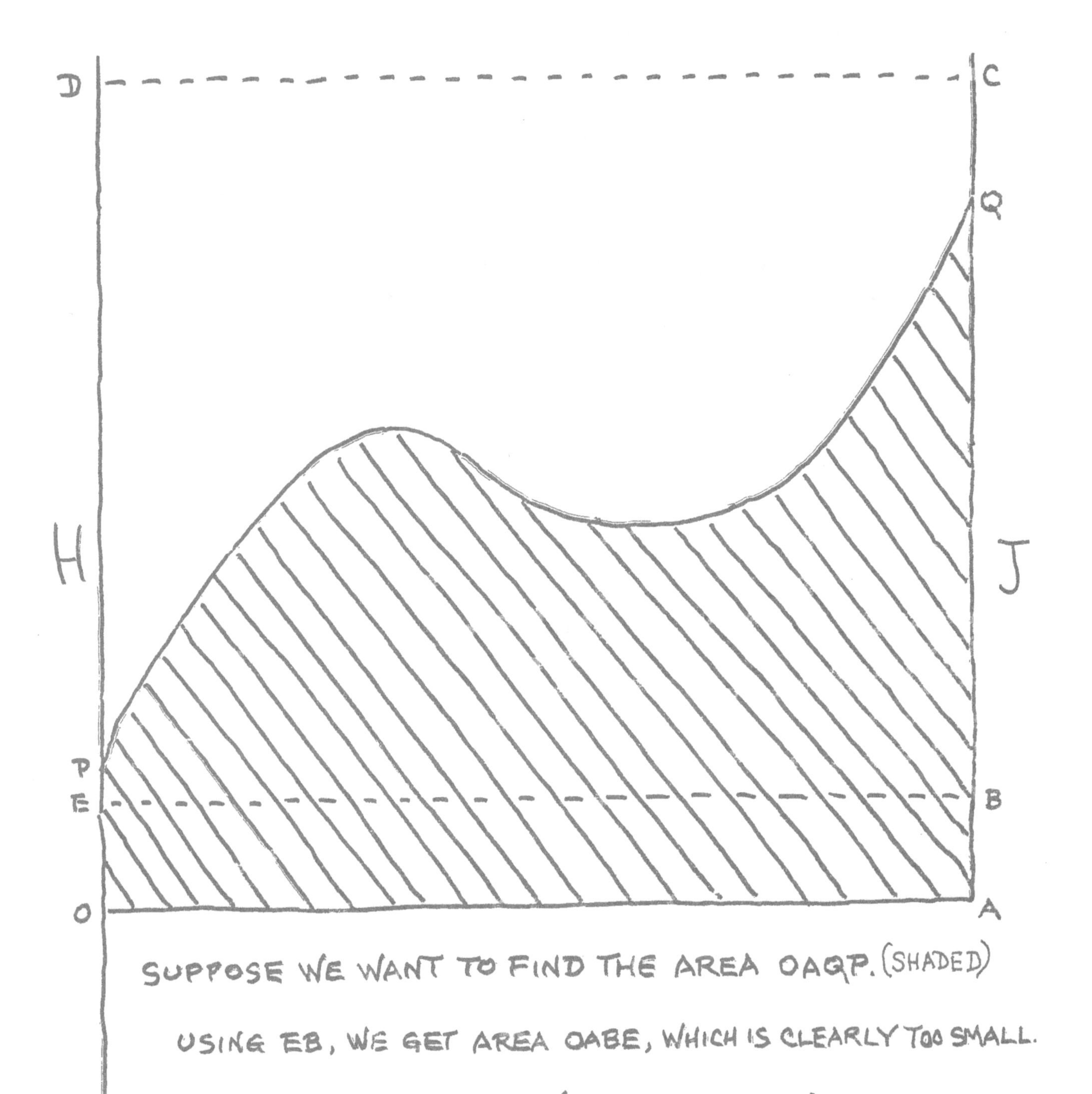

SUPPOSE WE WANT TO FIND THE AREA OAQP. (SHADED)

USING EB, WE GET AREA OABE, WHICH IS CLEARLY TOO SMALL.

WE SLIDE EB UPWARD (PARALLEL TO OA)
UNTIL IT REACHES DC.
THIS GIVES US AREA OACD, WHICH IS CLEARLY TOO LARGE.
BY CONTINUITY, THEREFORE, EB MUST PASS
THROUGH SOME MYSTERIOUS POSITION HJ,
WHICH IS JUST RIGHT, MAKING OAJH EQUAL TO OAQP.

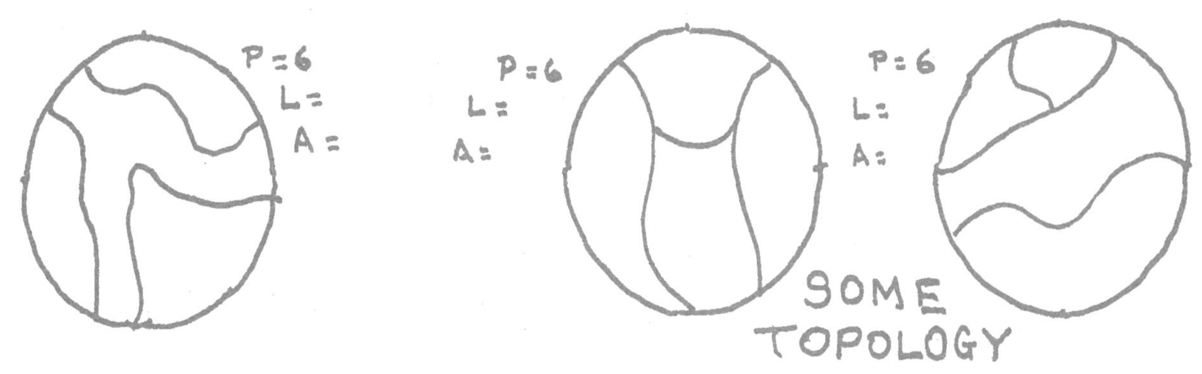

SOME TOPOLOGY

WE START WITH A CIRCLE:
INSIDE THE CIRCLE WE DRAW SOME LINES
(BEING CAREFUL NOT TO PUT ANY LOOSE DOTS
OR TO MAKE ANY 4-WAY INTERSECTIONS)
WE SHALL CALL A 3-WAY INTERSECTION A POINT.
TOPOLOGY IS NOT CONCERNED WITH THE MEASUREMENT
OF LINES, ANGLES, OR AREAS. WE SHALL CONSIDER MERELY
HOW MANY THERE ARE, AND HOW THEY ARE CONNECTED.

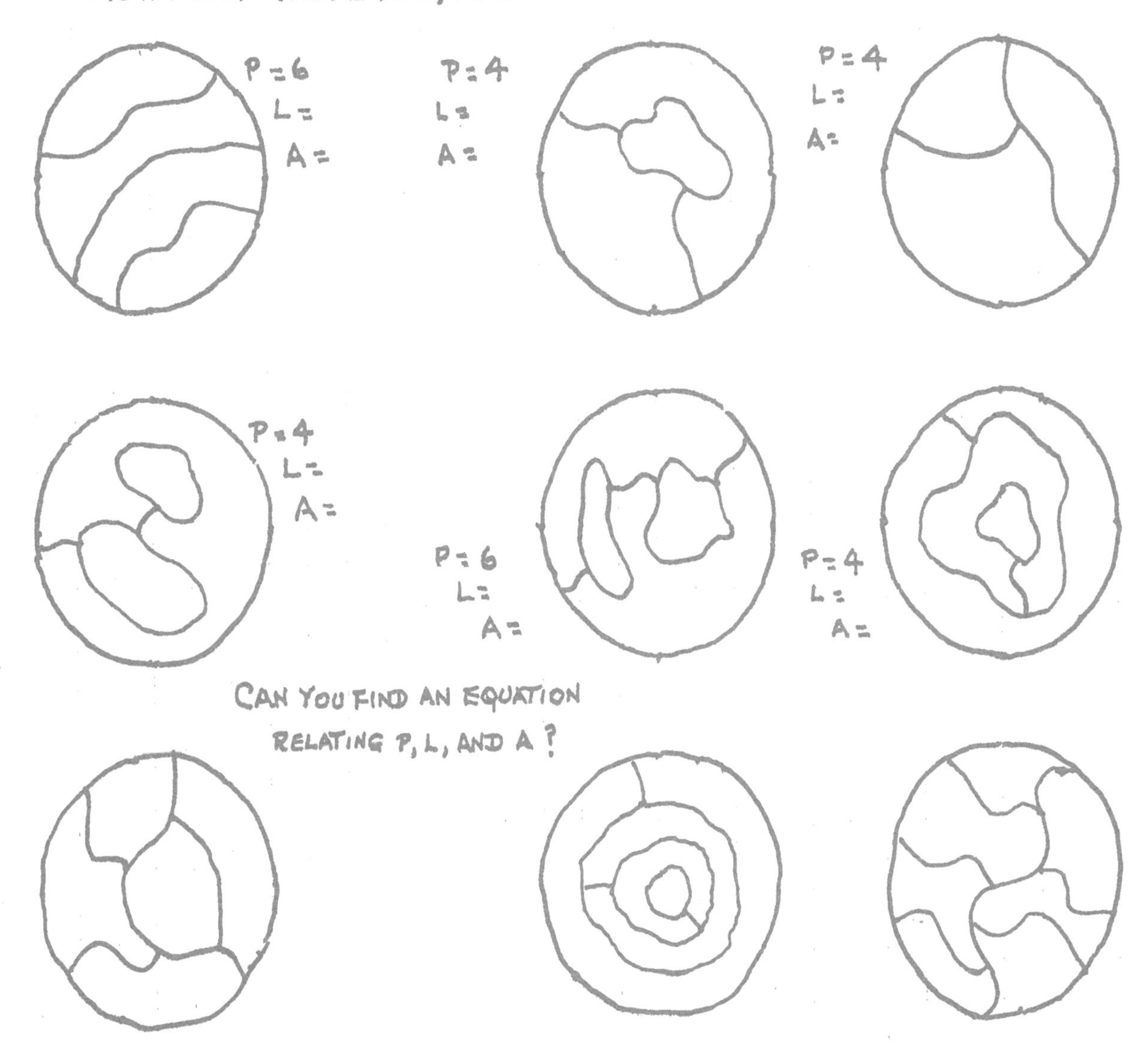

HERE ARE SEVERAL TOPOLOGICAL CREATURES:

IF WE DEFINE A TOPOSAUR AS A SET OF CONNECTED LINES
SUCH THAT EACH POINT IS CONNECTED TO EVERY OTHER POINT,
AND SUCH THAT EACH POINT
IS EITHER A 2-WAY INTERSECTION
OR A 3-WAY INTERSECTION,
THEN SOME OF THESE CREATURES FAIL TO QUALIFY:
(14, 19, 22)

THERE ARE 7 OF THESE: (, , , , , ,)

3 OF THESE (, ,)

2 OF THESE (,)

1 OF THESE

1 OF THESE

1 OF THESE

1 OF THESE

1 OF THESE

1 OF THESE

1 OF THESE

1 OF THESE

AND, OF COURSE #5, ABOVE.

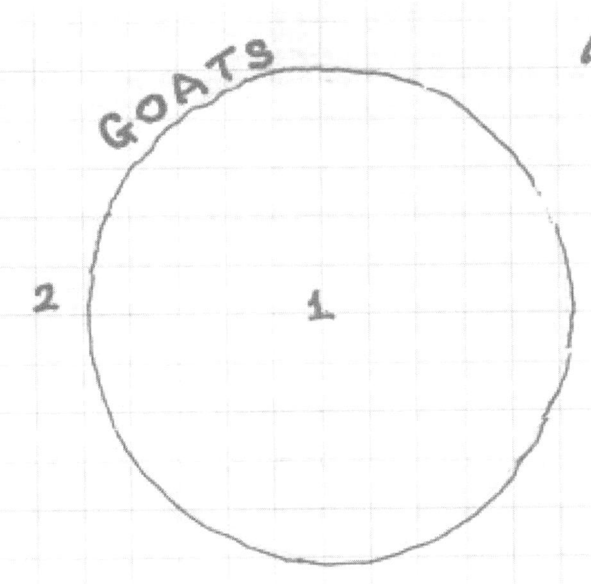

AS ALL THESE ANIMALS ENTER THE ROOM,
LET US CLASSIFY EACH ONE
ACCORDING TO WHETHER IT IS,
OR IS NOT, A GOAT.
IF IT IS A GOAT,
WE PUT IT INTO THE CIRCLE.
IF IT'S NOT A GOAT,
WE PUT IT OUTSIDE THE CIRCLE.

* * *

AN 83 POUND CASSOWARY
WOULD GO TO A PLACE
WHICH IS IN BOTH CIRCLES.
SO IT WOULD GO INTO REGION 2.
WHERE WOULD A 47 POUND
ANTEATER GO?
WHAT ABOUT A 17 POUND EEL?
WHAT ABOUT A 150 POUND CASSOWARY?

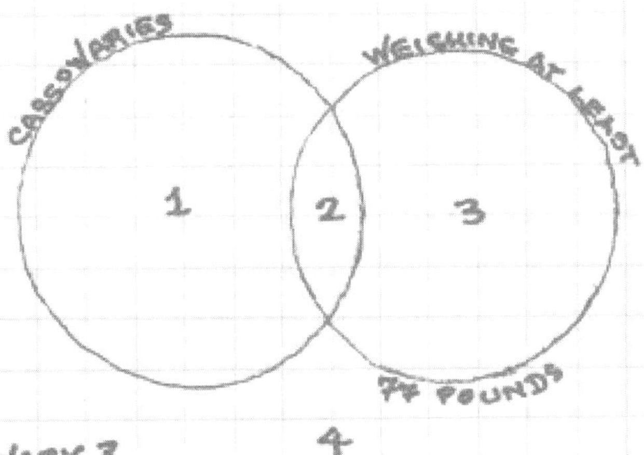

VENN DIAGRAMS

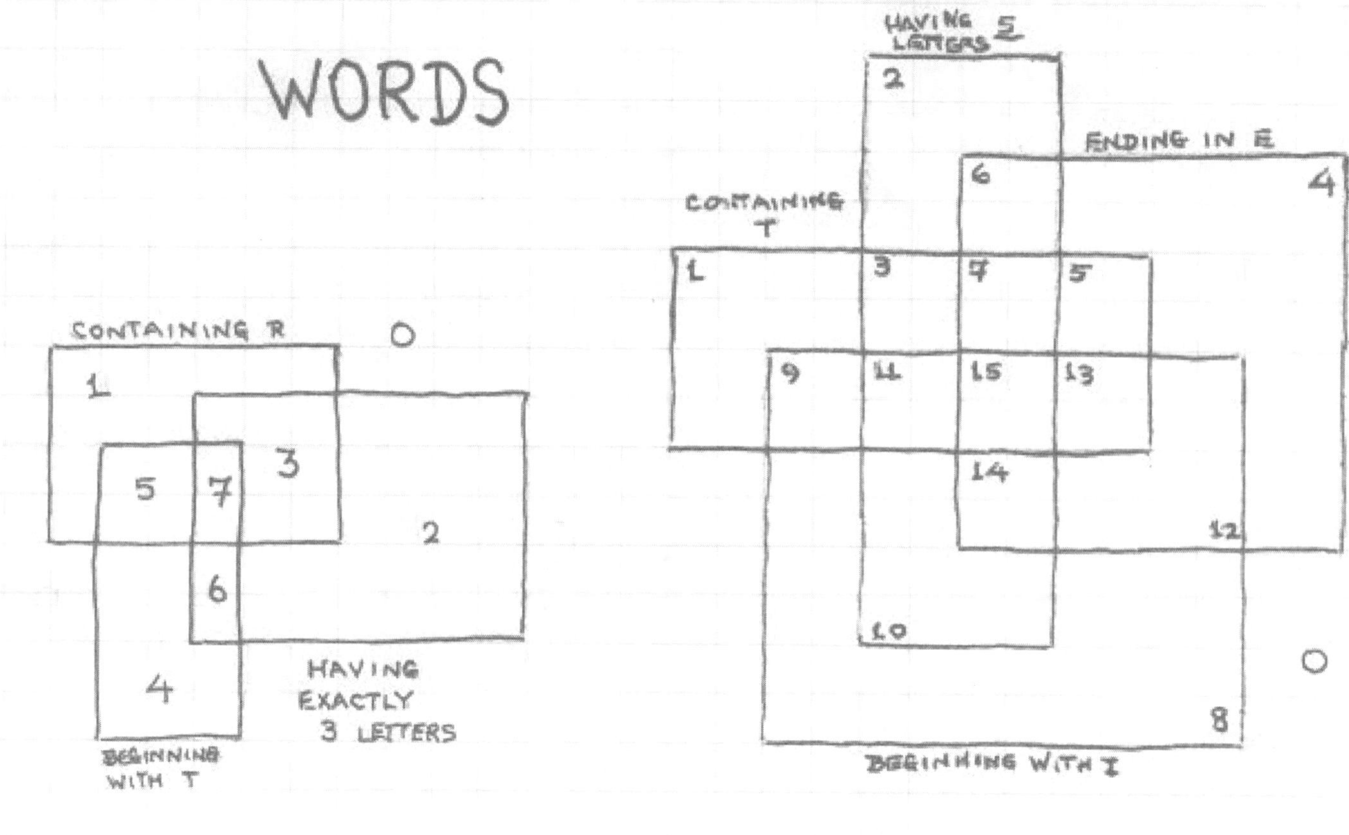

VENN DIAGRAMS

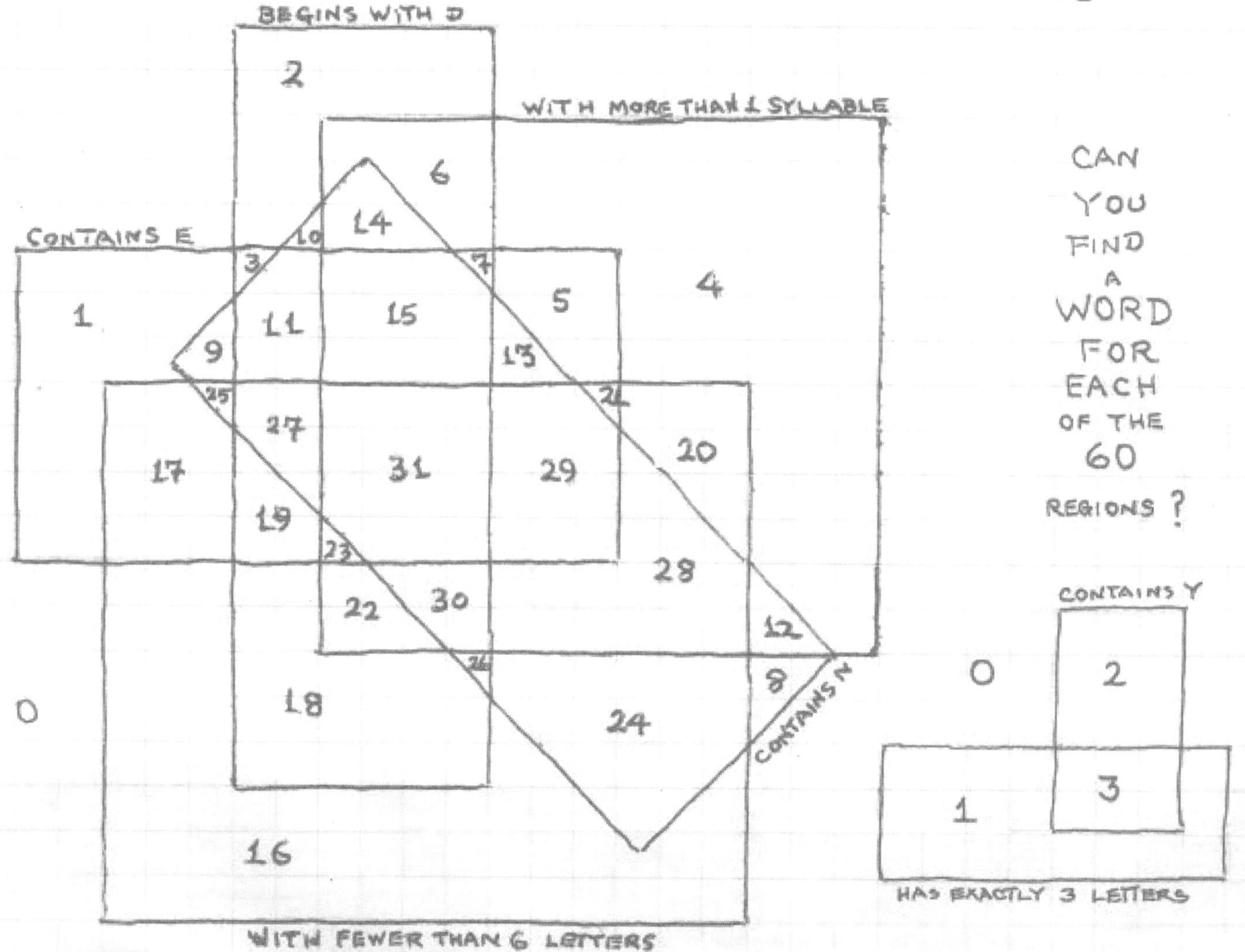

SHE WAS LOOKING BACK AT ME.

IS A CLEAR STATEMENT

I WAS LOOKING BACK TO SEE IF SHE WAS LOOKING BACK AT ME.

IS ALSO PRETTY CLEAR

SHE WAS LOOKING BACK TO SEE IF I WAS LOOKING BACK TO SEE IF SHE WAS LOOKING BACK AT ME.

IS NOT QUITE SO EASY, YET NOT DIFFICULT

NOW TRY THIS ONE:

I WAS LOOKING / SHE WAS LOOKING / I WAS LOOKING / SHE WAS LOOKING BACK AT ME.

THEY HAVE AN AGENT SPYING IN OUR COUNTRY.

BUT IF HE'S REALLY OUR AGENT PRETENDING TO BE THEIR AGENT SPYING IN OUR COUNTRY.?

THAT MAKES HIM A DOUBLE AGENT.

HE'S <u>THEIR</u> AGENT / <u>OUR</u> AGENT / THEIR AGENT SPYING IN OUR COUNTRY?

THAT MAKES HIM A TRIPLE AGENT.

QUESTION: IS A QUADRUPLE AGENT DIFFERENT FROM A DOUBLE AGENT?
IF SO, HOW?

YOU TOOK AN EXTRA COOKIE!

I SAW YOU TAKE THAT EXTRA COOKIE!!

YOU KNOW THAT I SAW YOU TAKE THAT EXTRA COOKIE!!!

I KNOW THAT YOU KNOW THAT I SAW YOU TAKE THAT EXTRA COOKIE!V

YOU KNOW THAT I KNOW THAT YOU KNOW THAT I SAW YOU TAKE THAT EXTRA COOKIE V